Puppy Socialization

WHAT IT IS
AND HOW TO DO IT

Marge Rogers and Eileen Anderson

PUPPY SOCIALIZATION
—— PROJECT ——

Bright Friends Productions

LITTLE ROCK, AR

Puppy Socialization: What It Is and How to Do It/ Marge Rogers and Eileen Anderson. —1st ed.

Paperback: ISBN 978-1-943634-09-5
E-book (all formats): ISBN 978-1-943634-08-8

Library of Congress Control Number: 2021911303

Bright Friends Productions
1818 N. Taylor Street, Ste. 8
#327
Little Rock AR 72207

In Praise of *Puppy Socialization: What It Is and How to Do It*

"This book is a great resource for any new pet parent. The information is based in the most up-to-date evidence-based science of behavior and learning. It is also presented in an upbeat and easy-to-understand manner, with a healthy dose of real-world stories that helps the reader connect with the authors. The checklists, outside resources referenced in the text, and even quizzes help ensure the reader has in-the-moment support while using this book to help their puppy learn about the world. There are creative examples on how to set up socialization exposure opportunities for puppies outside of the traditional public settings normally recommended. These are especially important for anyone raising "COVID puppies," people who are in areas where restrictions are still in place, or for any owner who is unable to go out and about regularly with their puppy or new dog." — **Sara L. Bennett, DVM, MS, Diplomate, American College of Veterinary Behaviorists North Carolina State University, College of Veterinary Medicine**

*****

"This is the most definitive resource on all things Socialization that you could ever ask for. Marge and Eileen, with their considerable experience, wisdom and humor, have given us a true gift—a one-stop-shopping book that offers essential information for first time puppy owners and seasoned hobbyists alike. Whether you are raising a puppy for sports or work, or simply for being an enjoyable companion that you can take anywhere, what you need is in here. I will be referring all students and friends to this book from now on. Beautifully done." — **Leslie McDevitt, MLA, CDBC, CPDT-KA, Level 2 TAGteach Certified, author of the *Control Unleashed* series of books**

*****

"Marge and Eileen's new book satisfied my love of learning. With equal parts compassion and expertise, their book offers accessible insight to a topic crucial to the well-being of dogs and their families. *Puppy Socialization: What It Is and How to Do It* is the perfect tool for anyone with a love for dogs and those who care about the human-animal bond. I can't wait to incorporate this resource into staff training at our veterinary hospital." — **Dr. Cathy Kreis, DVM, Medical Director, Appalachian New River Veterinary Associates**

*****

"Marge Roger's and Eileen Anderson's new book *Puppy Socialization: What It Is and How to Do It* had me at "Remember: it's a human tendency to want to show the world your puppy, but that is not the same as showing your puppy the world! Resist the temptation to let your puppy become a magnet for human attention. The last thing you want to do is let him get overwhelmed and frightened." Then, the book continued to deliver beautiful information and ideas throughout.

Puppy Socialization will teach people all they need to know about socializing their puppies so they develop into the very best possible versions of themselves. The guidance in the book offers great advice for novice puppy owners and detailed ideas for old pros at raising puppies. With clear instructions covering all kinds of scenarios along with more than 120 photos and videos, this book is practical, helpful, and fun to read.

I don't claim to be psychic, but I have a feeling this book is going to be read and recommended by heaps of dog professionals and dog lovers who will be eager to spread the word. It's the perfect resource for everyone who has anything to do with raising well-adjusted, happy puppies, and its publication is good news all around!

If more people read this book and do what it advises, there will be more happy, well-adjusted dogs on the planet, and that's about the greatest future I can imagine." — **Karen B. London, PhD, CAAB, CPDT-KA, author of *Treat Everyone Like a Dog: How a Dog Trainer's World View Can Improve Your Life***

*****

"Wondering how to effectively socialize your puppy? You've been told it's really important, right? Up until now, a comprehensive guide to this essential task has been lacking. Marge and Eileen have created a wonderful resource to help both novice and experienced dog owners alike accomplish socialization of your puppy or dog with confidence! They have done a wonderful service to canines and their owners by writing this book. The explanations and instructions use sound science based information and are broken down into manageable easy to read and understand steps. In addition, they've also included invaluable resources in the way of video and photo links to help the reader learn all about dog body language and see real-life training in action. These links will absolutely take your learning experience to the next level. Your puppy deserves the best start in life. This book allows you to provide it!" — **Alanna Lowry, DVM**

*****

Table of Contents

Note from the Authors 1

How to Access the Media and Supporting Materials Attached to This Book 3

    Videos 3

    Other Resources 3

Note on the COVID-19 Pandemic 5

Chapter 1. Socialization: Things to Know Before You Get Started 7

    Canine Body Language: The Missing Piece of the Socialization Puzzle 8

    Socialization Defined 10

    Dr. Ian Dunbar's Work on Socialization 11

    The Clock is Ticking: All about Sensitive Periods 12

    The Experts Agree: Don't Wait 13

    What Happens During the Sensitive Period for Socialization (SPS)? 16

    Fear and Fear Periods 17

    A Word About "Pandemic Puppies" 20

    Puppy Socialization Checklists: Guidelines, Not Rules 21

    Common Myths about Puppy Socialization 22

    Are Puppies Blank Slates? 26

    The Puppy Culture Program 27

Chapter 2. Pairing and Priorities 29

    New Things Predict Good Stuff 29

    What Happens If You Expose the Puppy to Things without Using Food or Fun? 36

    But They Don't Do This in Other Countries and I Never Did This Before and My Dogs Were Fine 38

    Identify Your Priorities 38

Pet vs. Sports/Competition Dog: Should My Socialization Goals Be Different? ..40

Chapter 3. How Puppies Learn and How Socialization Fits In 43

Classical Conditioning 43

Operant Conditioning 45

Habituation vs. Sensitization 48

Bank Accounts 49

Review: Use Food and Play to Form Positive Associations 50

Chapter 4. Understanding Dog Body Language 51

Body Language Examples of a "Relaxed and Happy" Dog 52

Look at the Whole Dog 60

Body Language of Fear and Anxiety in Dogs 62

Body Language Examples of a Fearful or Anxious Dog 65

Look at the Whole Dog (One More Time) 85

My Dog Is "Fine" 86

Body Language Comparisons 88

Back to Context 94

How to Greet a Dog: Greeting and Lifting Up Puppies 95

Take Our Quiz 101

Chapter 5. Socialization! Starting at Home 103

Creating Great Associations at Home 104

Building Your Skills 107

What You'll Need to Get Started 108

What You'll Do: Introducing Your Puppy to Novelty at Home 109

Video Examples of Building Positive Associations with Household Objects 110

Beyond the Basics: Introducing the Vacuum 115

A Word about Leashes 117

People 118

Puppies and Children 126

Planned Sound Exposures...135

Unexpected Sounds and Events 138

Handling 139

Other Dogs at Home..143

Spending Time Alone.. 148

Chapter 6. Socialization Away from Home....................**153**

When to Start Outside the Home 153

What You'll Need to Get Started 154

Riding in the Car 156

Planning Your Outings..158

Examples of Socialization Trips to Outdoor Environments.................164

Meeting People Out in the World 174

Planning Ahead for Indoor Environments 178

Example of a Socialization Trip to an Indoor Environment.................179

Objects Outside the Home 180

Sounds Outside the Home 185

Dogs from Outside the Household 188

Other Animals.. 191

The Veterinary Clinic 192

The Groomer 193

If Your Puppy Gets Overwhelmed: Planning for the Unexpected.................195

Puppy Classes 202

Chapter 7. Fear and Special Challenges **209**

Abnormal Fear During the SPS 210

Special Cases...215

Chapter 8. What's Next? **221**

Acknowledgements.. **223**

About the Authors **225**

About the Authors.. 227

Media Credits and Permissions...228

References 229

For the dogs and the people who love them

Disclaimer

The suggestions in this book are based on the best scientific information available. We have made every reasonable effort to present current and accurate information, but make no guarantees of any kind and cannot be held liable for incorrect or outdated information or the implementation of that information. The suggestions in this book can't take the place of a professional trainer who has eyes on your individual dog. The information about dogs' behavioral health should not be substituted for advice from a veterinarian. We are not liable for any damages or negative consequences from any treatment, action, application, or preparation performed by any person reading or following the information in this book. We are not responsible for any decisions other people make about their dogs.

We mention the names of products, businesses, and organizations that we have found helpful, but we do not assume responsibility or liability for them.

All service marks, product names, and trademarks mentioned are the property of their respective owners.

Note from the Authors

This book is a labor of love from the two of us: Marge Rogers and Eileen Anderson.

We wrote it because we want you to have the best relationship possible with your dog. We want you to be able to take your dog places, have visitors in your house, add new members to your family without stress, and do sports with your dog if you both enjoy that. We want your dog to feel safe in this human world and for you and your dog to enjoy life together. Socialization—the process of creating positive exposures to the world during a certain period of a puppy's development—is key to letting that happen. Also—here's a secret. Socialization done right will be fun and rewarding for you. It is a joy to see a puppy learn about the world, grow in confidence, and look forward to adventures with you.

Both of us grew up with dogs and later discovered the joys of training them. We learned some good information and some not-so-good information. We became passionate about sharing good information and dispelling myths, and both of us changed our careers as a result. We now do one-on-one training, give classes, create videos and webinars, and do lots of writing.

We know that people with widely differing experience with dogs will be reading this book. If you are brand-new to puppies and socialization (or it's been a while) but don't care as much about the *why* of socialization and want to get right to the nuts and bolts of *how*—that will work, but here is something very important. Yes, you may skip some of the why, but you **really should** read the chapter on dog body language. It's part of the "how." Study it, and start studying the dogs you see and know. This is because you have to learn how to read your puppy and to hear what he is saying to you to be able to socialize him properly. Not being able to understand how dogs communicate can cause your socialization attempts to crash and burn.

Sorry to be so blunt. But we've seen way too much crashing and burning, too much tragedy, all happening despite the best attempts of loving pet owners. Again, that's why we wrote the book.

If you love "all things dog" as we do, and have some recent experience with this topic, we hope you will bear with us. We explain the nuts and bolts for people who are new to socialization or even new to having a dog in the family. We still think you'll find some useful takeaways. Thank you for being here with us.

We have been collaborating on projects together for more than 10 years, and this book was born of that partnership. We both worked on every page of it. For most sections of the book, even we couldn't tell you which one of us originally wrote it. We both wrote and we both edited. Most of the time we write as "we," but if one of us is telling of a personal experience, we'll note that (e.g., "Marge here").

We hope you enjoy reading this book as much as we enjoyed writing it. And we hope you have a fantastic life with your puppy.

How to Access the Media and Supporting Materials Attached to This Book

Congratulations! You purchased more than a book. You purchased a multi-media resource!

Videos

This book includes access to more than 50 videos that are integral to the material we cover. We put all the videos onto a web page so they can be viewed as they occur in the book. You can find them at **https://PuppySocialization.com/videos**.

We've made it easy to tell when there is a video to watch. Images in the book with a caption and a "play" button on them indicate that a video accompanies and illustrates the text in the book. The videos in the book are numbered and captioned so you can easily find the corresponding video on the web page. Here is a preview of what that looks like.

Video 1.1 Four 3-week-old puppies tumble around

Please plan to watch the videos as you go along for the best learning experience. You'll need a live internet connection to do so.

Other Resources

We want to make sure you have all the information you need to socialize your puppy. We refer to many outside resources: position statements, books, posters, and more.

These are all compiled on a resource list on our website. The URL for that page is **https://PuppySocialization.com/resources**.

The video and resource pages are not included on the menus on our Puppy Socialization website, since the material on them is intended only for readers of this book. Be sure to open browser windows to the addresses whenever you read the book. We recommend you bookmark the pages so you can easily return to them.

Note on the COVID–19 Pandemic

As we publish this, many countries are still in the grips of the COVID-19 pandemic. People are still sheltering in place and may be for some time. We are both devastated at the impact the pandemic has had on our friends, families, communities, and the world. We don't know what the future holds, except that it appears that restrictions will vary greatly by time and location.

It may seem an odd time to publish a book on puppy socialization. There is so much human suffering and so much uncertainty. But what we do know is that some people are using the time at home to welcome a dog into their lives. With enforced isolation in many areas, some people have more time and seek the companionship of a puppy or dog. And as different as this time in our lives is, we can still make the most of the puppies' optimal period for socialization (we'll define that for you later). You can do so much at home to help your puppy adapt to a human world. For starters, see our chapter on socialization at home. And you'll be surprised at how many of the "on the road" exercises can still be performed even while shelter-in-place guidelines exist. In fact, social distancing is not necessarily a bad thing for lots of socialization exercises.

Your puppy may or may not be out of his socialization period when the crisis abates. Guess what? Our instructions still apply, regardless of the age of your puppy or dog. If your puppy is not in his optimal period for socialization, he will not progress as quickly. But that's normal for older puppies. Our guidance applies to dogs of all ages. As a matter of fact, not all the puppies shown in the videos are in their prime socialization period. That's okay; we don't stop those activities when puppies get older.

Most important, though, is that you stay safe. Follow federal, state, and local guidelines for your community. For example, just because there are sections in our book on taking your puppy to public locations, do not do so if this is prohibited by law or not safe. We include those sections so you have those resources available to you when it is permitted and appropriate.

If you have a puppy during the COVID-19 pandemic we are so happy for you. You have a bit of joy in these uncertain times. Thank you for inviting us along with you on this journey with your new puppy.

Chapter 1. Socialization: Things to Know Before You Get Started

So you are about to get your new puppy, or you just brought him home. Congratulations! We—Marge and Eileen—hope you have a wonderful experience with the newest member of your family.

You may have heard about socialization and have a general idea about it. Perhaps you've seen lists of things you are "supposed" to do with a new puppy. But we've found that a lot of people don't understand how crucial socialization is and don't know how to go about it so the pup has a positive experience.

We'll help you with that. In this book, we're not only going to explain what socialization is and how to do it, we're also going to show you, with real-life examples featuring other puppy owners like yourself. We have more than 60 videos linked in the book and most are exclusive. Many are short, even a minute or less, and demonstrate specific aspects of socialization. Some are longer and show you what progress can look like.

Sometimes things go exactly as planned. But, because these are real-life examples, sometimes things don't go as planned. We'll show you that, too, and explain what to do when the unexpected happens to you (because it will).

Our goals with this book are to teach you:

- what socialization is and how essential it is to your puppy's behavioral development (most people don't realize this!);
- how puppies and dogs learn so you can apply that knowledge to teaching your puppy about the world;
- about canine body language (your puppy's only language) and show you real-life examples;
- how you can (and should!) start socializing your puppy at home and away from home;
- what to do when things aren't going well, or if your puppy gets scared when you are socializing him; and
- the critical observation and physical skills you need to be able to effectively socialize your puppy.

We want to tell you up front that no book can take the place of help from a qualified, humane, force-free professional trainer. Most have worked with hundreds of puppies and can give you focused help for your individual puppy using non-frightening exercises and advice. They can give you a jump-start to a wonderful life with your puppy as you help him grow into a great dog. But we are realists. Not everyone can afford a trainer, especially in these uncertain economic times. So we are giving you all the information we can to help you successfully socialize your puppy. And a good book has advantages, too. You can take all the time you need to study the resources we show you about canine body language. We have examples of all sorts of breeds and breed mixes. In our many videos, you'll see real pet owners learning to do the things we are talking about. And you can learn to do them, too.

Please note that some situations with puppies, mostly those involving abnormal fear and aggression, require a professional. We'll describe those in the book, but this book can't give you all the help you need if you are in that situation. We hope that you'll do anything you can to get a qualified professional on board.

Canine Body Language: The Missing Piece of the Socialization Puzzle

There's something you need to know before you begin socializing your puppy. This "mission critical" information is not part of many owners' socialization plans. But if you miss this one, all your work may be in vain.

Before you begin doing this thing called socialization, it's critical to understand how puppies and dogs communicate through body language. You can't begin to introduce your puppy to his new life until you can tell when he is relaxed and happy or worried and avoidant. People try to socialize pups all the time without knowing the difference, and that's where things can go wrong (and boy can they!).

The pup in the adjacent photo is enjoying a socialization outing. We'll teach you how to tell whether or not your puppy is having a good time.

In our experience, most people think they can tell when a puppy is scared or worried. And it's true, they might be able to recognize the more obvious signs of fear. But many people are surprised to learn about dogs' more subtle signs of fear and worry, and how much their dogs are trying to tell them that they simply don't "hear." Without knowing the subtle signs, an owner can end up accidentally putting their puppy in a frightening or even traumatic situation while trying to show him the world because they missed his whispers for help.

Creating fear through exposure to new things is the opposite of socialization. It sets your puppy up to be scared of the world, and unfortunately, such fear can be persistent for the rest of the dog's life.

Being able to read dog body language is so important to your puppy's socialization that we've dedicated an entire chapter to it. Regardless of your dog or puppy's age, it is important to learn to recognize when he is even a little bit fearful or worried.

Is the young dog in the following photo relaxed or stressed? You'll learn to tell the difference.

We'll return to this photo later after you've learned what to look for.

Socialization Defined

The *Merriam-Webster.com Dictionary* defines socialization this way:

"Exposure of a young domestic animal (such as a kitten or puppy) to a variety of people, animals, and situations to minimize fear and aggression and promote friendliness" (Merriam-Webster, n.d.).

This definition is pretty accurate because it not only mentions what you should do (expose the puppy to different things) but also why you should do it (to minimize fear and promote friendliness).

But why does this work? Aren't all puppies born naturally friendly? How can exposing the puppy to different things minimize fear? How can it promote friendliness? Those questions are what we will answer. But we want to tell you right up front that the way to create positive exposures for your puppy is to pair those experiences with something the puppy loves, usually food and play. Our goal is to make the new thing (environment, person, object, noise) a predictor of great stuff for your puppy.

The goal of socialization isn't merely exposing your puppy to new things. The goal is to help your puppy form positive associations, using food and play, with the things he's going to encounter during the course of his lifetime. We want to help you teach your puppy to love the rest of the world as much as he loves food and play.

Socialization is an investment in giving him a happy life and a comfortable place in your family.

Dr. Ian Dunbar's Work on Socialization

Dr. Ian Dunbar, in the 1980s, was one of the first to introduce information about puppy socialization and training to the general public. Before the work of this pioneering veterinarian, it was often recommended not to even start training your puppy until he was 6 months of age. This recommendation was due to two factors. First, the methods at the time were based primarily on older, dominance-based training models with choke chains commonly used. They were rightly viewed as too harsh for young puppies. (Those training models and tools are unnecessary for grown dogs, as well.) Second, people were worried about disease transmission. We know now that puppies are protected by maternal antibodies and there is little risk of disease transmission in a well-run puppy class (Korbelik et al., 2011).

Dr. Dunbar revolutionized puppy training by bringing puppy behavioral development and food-reward training to the public. He started his revolutionary Sirius Puppy Training Classes in the 1980s in the San Francisco Bay Area. He published his manuals *Before You Get Your Puppy* and *After You Get Your Puppy* in 2001 (Dunbar, 2001b; Dunbar, 2001a).

Dr. Dunbar's free-downloadable book, *After You Get Your Puppy,* was groundbreaking. It contains valuable information about training and socialization. I (Marge) still consult the book and usually pick up a new tip in the first five minutes of reading, and I've been doing this for a while now. At the same time, I've learned over the years of working with clients that his advice at times can be rather daunting. Yes, it would be great if your puppy got to meet dozens of individual humans during his socialization period, but **only if he is building good associations.** Exposing a scared puppy to dozens of people will likely create a fearful adult dog.

So yes, we want you to provide as many, varied experiences as you can to your puppy. But the **quality of the experience** is what will help your puppy the most, not raw numbers of exposures. That's why we'll return again and again to reading body language.

The Clock is Ticking: All about Sensitive Periods

The behavioral development of dogs has been well studied. What we've learned is that we have a limited time to effectively expose puppies to the things they are likely to encounter during the course of their lifetime. Mother Nature has an open "window" during which puppies are primed to be more open to accept novelty and new experiences. The common name for this crucial period in their development is the *socialization period*.

Scientists across several disciplines refer to this window and other windows in an animal's development as *sensitive periods* (Bateson, 1979; Knudsen, 2004; Montessori & Carter, 1936).

Veterinary specialist in behavior Dr. Karen Overall explains sensitive periods and their importance well:

> *"A sensitive period is best defined as a period when animals can best benefit from exposure to certain stimuli, and if deprived of such exposure, there is an increased risk of developing problems attendant with the stimulus"* (Overall, *2013, p. 124).*

Besides the one for socialization, there are also sensitive periods for the development of vision and many other aspects of physiology and behavior (Fox et al., 1968).

Since there are different types of sensitive periods, we'll refer to this special time in puppies' lives as the *sensitive period for socialization*, or *SPS*. The SPS runs approximately from age 3 weeks, when puppies develop vision and hearing, through about 12 weeks (Overall, 2013, p. 123). This sensitive period is a magical time. The puppy's perception of the world is more plastic—moldable, if you will. The puppy's experiences during the SPS, or lack of experiences, will likely impact him for the rest of his life.

It is also common to refer to *primary* and *secondary* sensitive periods for socialization (Martin & Martin, 2011, p. 26–27). In this method of classification, the primary period is between 3 and 5 weeks. During this time, puppies are learning to use their senses and to interact with their peers and their mother. Then, the secondary sensitive period is from about 6 to 12 weeks. This is the period we discuss most often in this book—the time when puppies are learning about the rest of their world.

You'll sometimes see the SPS called the *critical period* (Colombo, 1982), but most scientists are now using the term sensitive period. By any name, this period of a young animal's development is crucial for certain physiological changes or social behaviors to develop.

In the following video, you can see a litter of 3-week-old Portuguese water dog puppies. They sure are cute. But maybe we are biased: one of those puppies went home to live with Marge. The puppy to the left of your screen (on top of the other puppy) later became known as Zip. You'll see him throughout this book. At 3 weeks old, these puppies were likely entering their SPS.

This video, and all the videos featured in the book, can be found at https://PuppySocialization.com/videos. Please open a web browser to that page and bookmark it so you can easily watch the videos that are integral to the book.

Video 1.1 Four 3-week-old puppies tumble around

The Experts Agree: Don't Wait

See the mat under the puppy in the adjacent photo? Can you guess why it's there?

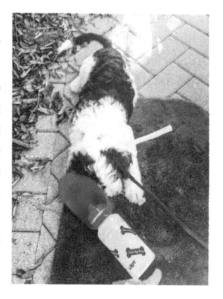

Using a mat is one of many things we can do to protect puppies from disease when we take them out and about before their vaccinations are complete. Not only can it keep very young puppies completely off the ground, but it can also continue to function as a "home base" on outings when you need to limit your pup's roaming around.

For many years, socializing puppies and keeping them safe from infectious diseases were seen to be at odds. It was often recommended to sequester puppies at home until they had all their vaccinations. This is no longer the case. The experts in the field have weighed in, and they have shown that socialization is too important to delay.

Dog trainers (see PuppySocialization.com/resources for help on finding a good trainer or behavior expert) and behavior consultants have figured out ways to get pups out in the world while taking steps to protect them from infectious diseases. They recommend following the guidelines of the American Veterinary Society of Animal Behavior (AVSAB) and **start taking the puppy out for careful socialization trips 7 days after his first set of vaccinations** (American Veterinary Society of Animal Behavior, 2008a). We link to this document at PuppySocialization.com/resources.

It's true that there are dangers out there to a young puppy's immune system. While we were writing this book, there was an outbreak of parvovirus in Marge's area. Marge, and hopefully the other trainers in her area, were able to tell their clients how to deal with this tricky situation.

The one thing you can't do is decide not to socialize. Trainers have become even more creative now, since COVID-19 arrived and changed so many aspects of daily life.

The AVSAB position statement on puppy socialization is worth checking out. It opens with, "The primary and most important time for puppy socialization is the first 3 months of life." It also includes that "behavioral issues, not infectious diseases, are the number one cause of death for dogs under 3 years of age" (American Veterinary Society of Animal Behavior, 2008a).

That's surprising news to many people. How can "behavioral issues" be deadly? There is a logical, tragic progression. Puppies who are kept in isolation during their SPS are more likely to be fearful. Fearful dogs are more likely to bite (Borchelt, 1983). And dogs who bite are at risk of losing their homes or their lives (Siracusa et al., 2017).

Now you can see why socialization is critical.

Veterinary behaviorist Dr. Lore Haug puts it like this:

"Socialization deficits are arguably the most prominent factor in the development of aggression in physiologically normal dogs. . . . Deficits in social interaction may become more problematic as the animal matures and neophobia and competitive interactions become more salient. Mere exposure to other people and dogs is not sufficient to guarantee adequate social skills. Interactions must be monitored to ensure that the puppy has a positive and enriching experience" (Haug, 2008).

This is why lack of socialization is dangerous to a dog's behavioral health and safety.

You still might hear the old-fashioned advice to keep the puppy sequestered at home. Some veterinarians still suggest keeping your puppy home until he's had all his shots at 16 weeks of age. We appreciate those veterinarians. They are taking the responsibility for your pet's health and wellness seriously. Their goal is to reduce health risks for your puppy and keep him well. For years this was standard advice, but that was before the intense study of the sensitive periods in the last few decades. Keeping him home misses a big part of your puppy's overall health: his behavioral wellness, which is critical.

The good news is that you can socialize your puppy *and* prevent exposure to contagious diseases. By taking some care and following the steps we describe in chapter 5 and chapter 6, you can socialize him safely.

Who Are These Experts?

So far in this book we have mentioned several types of professionals with expertise in dog behavior. Here is a quick overview of the professionals we recommend. Unfortunately, dog training and behavior modification are unregulated fields, and anyone can hang out a shingle and call themselves a trainer or behaviorist. We want to help you find the right professionals, should you need help with your puppy's (or grown dog's) behavior.

Here are the credentials to look for in the United States.

- **Certified behavior consultants.** These professionals are certified by the Certification Council for Professional Dog Trainers, the International Association of Animal Behavior Consultants, or the Pet Professional Accreditation Board. Marge is a behavior consultant certified by CCPDT.
- **Certified applied animal behaviorists.** These professionals are certified by the Animal Behavior Society.
- **Veterinarians.** These professionals are graduates of an accredited university veterinary program and are licensed through the North American Veterinary Licensing Examination and the state in which they practice.
- **Board certified veterinary behaviorists.** These professionals are veterinarians with additional specialty training in behavior and are certified by the American College of Veterinary Behaviorists.

In this book, we use the more general term *behavior consultant* to also include certified applied animal behaviorists. We don't ever use the term behaviorist by itself because unfortunately, many people without appropriate education or credentials use it to describe themselves. The only credentialed behaviorists in the United States are certified applied animal behaviorists and veterinary behaviorists.

All of the accrediting bodies listed above have directories you can consult both to find an appropriate professional or to check the credentials of one you are considering. We link to an article by the American Kennel Club on our resource list (PuppySocialization.com/resources) that reviews the professional categories listed above in more detail.

What Happens During the Sensitive Period for Socialization (SPS)?

As we mentioned, the duration of the SPS in puppies is approximately from 3 to 12 weeks of age. During the early part of the SPS, when puppies are about 3–5 weeks of age, they haven't yet developed a fear response, such as a fear of new or novel items (Scott & Fuller, 2012, Ch. 5 The Critical Period, "Boundaries of the Critical Period"). Since everything is novel to brand-new puppies, if they were afraid of novelty, they would be afraid all the time. So selection by nature and by humans has favored pups who are not fearful when they are very young. Their brains are like little sponges when they are babies.

But soon after, about when the puppy is 5 to 6 weeks old, the potential for fear starts creeping in (Coppinger & Coppinger, 2002, p. 115) (Scott & Fuller, 2012, Ch. 4 The Development of Behavior, "Period of Socialization"). This is an approximation; most experts agree there are breed and individual differences (Serpell, 2017, Section 6.2.4). Before this change, puppies have a reflexive startle response to sudden changes in the environment, but don't display behaviors associated with fear.

There may also be a spike in fear around 8 weeks of age (Serpell, 2017, Section 6.4.2). We discuss this in more detail in the section on fear periods below. Then, at around 12 to 16 weeks of age, the sensitive period ends, and with it the best chance of getting a puppy comfortable in the world. It makes sense that it would be a survival advantage for tiny puppies to accept that the experiences and things they encounter are part of their world. But as they grow older, it becomes advantageous to be wary of new things because they could be dangerous. If a dog has not been exposed to much of the world before the socialization window closes, his natural response to something he hasn't encountered before is usually fear.

For example, Dr. Karen Overall describes the outcome of a socialization study:

> *"Pups that were kept in kennels beyond 14 weeks of age were very timid and demonstrated a lack of confidence in any circumstance other than the kennel. These dogs would not voluntarily leave the kennel and became truly phobic of anything novel (neophobia)" (Overall, 2013, p. 123).*

Unfortunately, dog trainers and behavior consultants see puppies like this all the time. They are afraid of anything new—as small a difference as a plant being moved from one part of the room to another (that's a real client example from Marge). Or an empty trash can the puppy walks past daily that has yard waste in it one day (also a real example).

As Dr. Overall puts it,

> *"Dogs not allowed to explore new environments by 14 weeks will not voluntarily do so. If forced to do so, they freeze and become extremely distressed" (Overall, 2013, p. 127).*

It is vital to understand what it's like to live with an unsocialized puppy as he grows into an adult. You will probably spend considerable time and resources helping him feel safe in his own home and around new people. Family visiting for the holidays, a child's friend visiting, delivery drivers—every person may be a challenge. **If he is afraid and you don't intervene, he will not grow out of it.**

As a matter of fact, the behavior often intensifies as the puppy gets older. His fears of people, dogs, environments, or new things will likely impact whether you have friends and family over to your house, whether walking your dog is relaxing or stressful, and a host of other things impossible to predict. Living with a fearful dog changes your life.

Fear and Fear Periods

Let's talk about fear for a moment. Although a major goal of socialization is to help a dog not to be afraid of lots of things, we need to realize that fear is a natural and necessary part of life.

Veterinary behaviorist Dr. Wailani Sung defines fear as follows:

"Fear is the instinctual feeling of apprehension caused by a situation, person, or object that presents an external threat—whether it's real or perceived" (Sung, 2019).

Dr. Sung described the internal experience of fear. There are also typical *behavioral* responses to fear. Veterinary behaviorist Dr. Christine Calder describes three categories of responses here:

"Dogs have three basic strategies they may choose to use when they are afraid or anxious: fight, flight, or freeze. These behaviors that accompany fight, flight or freeze are all normal social behaviors in dogs" (Calder, 2020).

Some experts add a fourth category: "fidget or fool around."

There are countless behaviors in dogs that fall into the fear response categories. We will describe and show you many of them in our chapter on body language.

Fear is not, in itself, a bad thing. Fear helps us protect ourselves. It is natural and functional to be startled by a loud noise. It means something big just happened and we may need to get out of the area!

What we want to prevent, with socialization, is a dog who is afraid of his world. As Dr. Sung put it above, reacting to threats that are perceived, not actual. We want him to be comfortable with things he'll encounter regularly like hearing a garbage truck rumbling by, seeing the veterinarian, and having his nails trimmed. How many dogs freak out the first time they see a man with a beard, hat, or sunglasses? Men with these attributes are not necessarily dangerous, and we would rather our dogs take them in stride. But puppies need to be exposed to them during the SPS for that to happen.

For his own sake, we would never want a dog to live in fear. But also keep in mind what we mentioned before: fearful dogs are also more likely to bite (Borchelt, 1983).

When you get your 8–10-week-old puppy, he is probably still pretty spongy. He's primed to accept his new world. He is flexible enough to accept this enormous life change, where all he knew is gone and he has a new family. But he is not free from fear. If he were totally a sponge, you could take him everywhere and show him the world and that'd be that.

Many people learn the hard way that this doesn't work. He can and will be afraid of things, and you can't proceed as if it isn't happening. **You have to ensure that his**

exposures to the important things in his world are positive and fun and not overwhelming. That is the heart of socialization. That's what we are going to show you how to do in this book.

The puppy in this photo is lovely, isn't he? It might not be obvious, but he is afraid. We'll teach you to recognize that later. But since we're talking about being afraid, it's important to know there are times when puppies are more likely to be fearful.

Fear *Periods*, Too?

We're sorry to throw one more term at you, but it's important, and you need to know how it fits in with all this stuff about sensitive and critical periods. The term is *fear period*. We mentioned earlier that puppies develop the ability to have a fearful reaction at about 5–6 weeks old. But that isn't a fear period. It just means their neurological system has developed the capability of having a fear response.

A fear period is a time in which puppies are particularly vulnerable to fear. And there is often one right smack in the middle of the SPS. As we mentioned above, it may occur right around 8 weeks of age. Note that this is often the time you are taking your puppy home. It is often observed by owners, trainers, and breeders, and is supported by a study done with beagles. In this study, the pups were particularly vulnerable to fear of humans during a period that started at the 8-week mark (Serpell, 2017, Section 6.4.2). It may be that puppies from about 5–8 weeks old don't retain fear as well and bounce back more easily, and puppies 12 weeks and older have bonded with humans and become more stable. But right in between, and right about when many people get their pups, is a period where puppies can get scared easily and perhaps not bounce back as well (Fox & Stelzner, 1966).

Breeds and individual dogs will respond differently (Serpell, 2017, Section 6.2.4), but since a pup is likely to be more vulnerable to fear from about 8–10 weeks old, be extra careful when you first bring your puppy home.

Sometimes if something scary or traumatic occurs, it can affect an animal for the rest of its life. This is referred to as *single-event learning*. It can happen anytime, but since puppies are already more prone to fear at around 8 weeks of age, they are likely to be extra vulnerable at this time.

Many veterinarians, dog trainers and other professionals have observed what appears to be another fear period that occurs during canine adolescence (Martin & Martin, 2011, p. 28), but so far it hasn't been shown in a scientific study. Some observation of captive wolves seems to show a fear period at around 4–5 months old (Serpell, 2017, Section 6.4.2). Dogs aren't wolves, so we mention this only because you may hear talk about secondary fear periods in dogs. But understand that secondary fear periods have not yet been validated in dogs by a scientific study.

Whether or not there are multiple periods, the take-home message is to always observe your puppy carefully. And that's exactly what we will teach you to do in this book!

A Word About "Pandemic Puppies"

As we mentioned, many people have welcomed dogs and puppies into their homes since the beginning of the COVID-19 pandemic. That's likely to continue. I (Marge) have observed an alarming trend with these puppies that are spending extra amounts of time at home. While they keep their owners company at home, many of the puppies I've encountered are learning very little about the larger world. They are atypically afraid of new people (even ones that don't seem scary), new locations, and novel items like a new toy or a different water dish. Some of these puppies are so fearful that the situation is an emergency.

But even during a pandemic, this doesn't have to happen. You can do so much right now to influence your puppy's behavioral development. All puppy owners need the information in this book, but owners with puppies during the COVID-19 pandemic *really* need it. The good news is that you can do a lot of things at home, starting right now. You can teach your puppy about novelty, noises, people, and sudden changes in the environment. If you are welcoming a new puppy during the pandemic, we're especially glad you're here.

Puppy Socialization Checklists: Guidelines, Not Rules

Some noteworthy veterinarians, trainers and organizations have created puppy socialization checklists. These are extremely detailed lists of people, animals, and objects that specify different physical characteristics. For instance, a typical list won't have an entry for "men." It will have multiple entries. Tall men. Men with deep voices. Men with beards. Men wearing hats. And lots more.

The world is full of variety and we want to make sure we include variety in our socialization plans. Puppies and dogs are great at noticing differences. So we want to make sure they experience plenty of "differences" or variety within each category. A man with a beard and wearing a hat still fits in the "men" category. Since puppies and dogs don't generalize well, we want them to learn about variety within categories of things.

Marge here. I think puppy socialization checklists are great. They give you ideas on how puppies see the world: with lots of distinctions. But here's the thing: the lists are guidelines, not rules.

My husband and I have an ongoing discussion. I'm a "rows and columns" kind of person and trainer. Much like puppies, I like structure. I follow the rules. My husband is always telling me that the "rules" are just guidelines. Don't tell him I said this, but when it comes to puppy socialization, he's right. Strict adherence to a list with checkboxes can lead you to ignore things that are crucial.

So if you have a puppy, Tuesday doesn't mean you go on a scavenger hunt to expose your puppy to a man with a beard, wearing a hat, who is tall and loud and wearing sunglasses, just because that's the next checkbox on someone's list. If you are fixated on finding that one guy, you may miss a great opportunity to watch, from a safe distance, a woman mowing a lawn or a child on a scooter.

And just as important: Socialization is not merely exposing your puppy to those things on the checklist. It's the process of creating positive associations with those things. If you have a puppy who is a little worried around men and you are searching for men of all varieties everywhere so you can check them off your list, you could end up making your puppy more afraid of men.

Do use the checklists, but use them to get ideas. Don't use them merely as a form of socialization bingo or scavenger hunt. The goal is not just to check items off the list. The goal is to make sure your puppy is forming positive associations with those things. We link to some in the resource list for this book (PuppySocialization.com/resources) that we have created on our website. Use the lists to get ideas of the types of things, places, and people you will want to introduce your puppy to during his SPS.

Common Myths about Puppy Socialization

The outdated information and myths that have developed around puppy socialization are often actively harmful to puppies. These myths are one of the big reasons we wrote this book.

I (Marge) often work with distraught puppy owners who tried to do everything right. They followed the advice of a breeder, trainer, or friend—someone with the best of intentions—and their puppies did not thrive behaviorally. Sometimes they were told their puppy just had to "get over it" or "get used to it." And their puppies got worse.

It's more important that you remember what *to* do, but because socialization myths are so prevalent, we're going to list some of them. We have to, because you'll probably encounter well-meaning people who try to convince you of them. Please resist the temptation. I see the harm done by these myths almost every day.

Myth: Keep your puppy home until he's had all his shots. I (Marge again) hear this all the time from clients with fearful puppies. My clients heard it from a friend, breeder, rescue group, the receptionist or technician at the animal hospital, or their veterinarian. But veterinarians who specialize in animal behavior and the American Animal Hospital Association recommend that puppies receive socialization *before* they have completed their vaccinations. The professional society for veterinary behaviorists suggests that puppies may begin careful exposures **seven days after their first set of vaccines.** We go into this in more detail in "When to Start Outside the Home" below.

Myth: Take your puppy everywhere. Socialization is not about taking your puppy places so he merely "gets out." That can backfire badly and he could end up more fearful. For example, if you take your puppy to the farmers market or an outdoor art show so he "gets out" while you walk around and shop, you won't notice whether he is happily exploring and meeting people or is overwhelmed. You'll be focused on your own business; that's just human nature. You probably will have missed opportunities to pair his new experiences with food and play. (More about how to do in the section below: "What You'll Do: Introducing Your Puppy to Novelty at Home.")

Recently, I met a client with a 24-week-old puppy we'll call Sophie. Shortly after Sophie was adopted, the 6-week-old puppy traveled across the country on a plane with her new owner. Sophie met and was held by many, many people during her trip to a metropolitan area. She went from a rescue in a quiet rural area where she had not met many people or experienced many things to a noisy city full of people, vehicles, and activity.

Those situations were overwhelming for Sophie. While she didn't seem too scared to her owner, she was often held and not able to move away. The owner took her everywhere. While her owner's intentions were good, the missing piece was observing Sophie's body language to see how she felt during these interactions. She was likely petrified during many of those experiences. Now, at 24 weeks, Sophie is a fearful dog. She is afraid of new people. She barks at people she sees on the street when she is on leash. She barks at bicycles, children, and more when riding in the car.

I get clients with dogs like Sophie all the time and it breaks my heart. These owners tried to do the right thing. They followed well-intentioned but misguided advice from someone they considered an authority figure. Some even tell me they could tell their puppy was scared or that they were making him worse, but they were told by many sources to keep doing it. This myth is dangerous. Let's put it to rest.

Myth: Have him meet 100 people in his socialization period and then you're done with socialization. This is well-intended advice. By now we've covered some of the problems with it. But what do the behavioral experts say? The behavior management guidelines released by the American Animal Hospital Association highlight the risk of the "100 people and done" approach. The guidelines say,

> *"Either the presence or development of fear during sensitive periods is aggravated by forced social exposure. Overexposure can make fearful dogs worse, creating a behavioral emergency. Clients should be advised that any dog either beginning to withdraw from interactions or exhibiting outright fear should not have more exposure unless recommended by their veterinarian. If the behavior is extreme, a veterinary behavior specialist should be consulted. Continuing to expose fearful puppies in the guise of 'socializing them' instead sensitizes them" (Hammerle et al., 2015).*

A number set as a goal, like "meet 100 people," overlooks the most important thing: socialization has to be positive and tailored to each puppy. Additionally, your job is not done after your puppy is no longer in his SPS. You need to continue his exposure to the world through the first year of his life (Donaldson, 1996, p. 61).

Myth: If your puppy worries about strangers, have them feed the puppy treats. If the pup is already fearful, this practice will likely backfire. Forcing puppies to be around things that scare them in the hopes that they "get used to it" can make them worse. We know better now. We can do better now.

Myth: You can "socialize" your puppy by meeting unknown dogs on walks. First, we know socialization is more than mere exposure. Second, unknown dogs could injure your puppy or carry disease. And even if you know the dog, they could easily do something to scare your puppy. Interactions with adult dogs need to be planned with dogs you know are puppy friendly and up-to-date on their vaccinations.

Myth: You must take your puppy to puppy class. Maybe, but it depends on your puppy and the puppy class. We've observed wonderful puppy classes. And at the other end of the spectrum, we've seen puppy classes that were free-for-alls. Puppies bullying other puppies, puppies hiding, puppies "shut down" (which means they were too scared to do much), puppies stalking other puppies—all in the guise of "puppy class." Those kinds of classes can harm your puppy and can be worse than no class at all. We've also observed puppy classes where the main goal is to teach behaviors like sit, down, and come. Those puppy classes are missing the mark for puppies in the sensitive period for socialization. See our extensive section on puppy classes for more details. Choose wisely. Your puppy's behavioral health depends on it.

Because of the COVID-19 pandemic, some puppy classes have ceased, and many are being modified. Make sure you can keep both yourself and your puppy safe in any class and venue you consider. Many classes (and private puppy training) are now offered online and could be good options.

Myth: You only have to socialize your puppy to people and dogs. It's true that if a person does only these two things, they might be ahead of most owners. At the same time, it doesn't mean they'll have a well-adjusted dog. There are so many other important things. Did the puppy have fun in new places? Build positive associations with alone time? Have good experiences with children and a variety of people outside the home? Form positive associations with holiday decorations?

Animals living in a human world have so many things to get used to. People and dogs are only a small part of that world.

Myth: Dogs who are afraid of men or brooms or vacuums or (you name it) have been abused. At first glance, this item may not look like it fits here. But it touches on a very important myth: the idea that all fearful dogs have been abused. There is a more likely reason for their fears. Lack of appropriate exposure during the SPS usually manifests as fear later on. But whether your dog is worried about the broom because someone hit him with one or (more likely) he's never seen one in use before, the takeaway is that you can take action. Your dog doesn't haven't to spend his life afraid of brooms. That's why we need to work hard to carefully introduce puppies to the things they'll encounter during the course of their lifetime. Because it's much easier to teach puppies that brooms (and vacuums, and people, and skateboards) are unremarkable parts of their world than it is to teach an adult dog that these things mean him no harm.

Myth: If you already have a dog or dogs when you bring your puppy home, he will be "fine" with all dogs. Unfortunately, exposure to the dogs in his new family during the SPS is not enough to help him generalize his experience to other dogs. I (still Marge!) hear this a lot. "But he is fine with our dogs." Those other dogs in your household became part of your puppy's world when you brought him home. Your puppy still needs to meet other puppy-friendly, vaccinated dogs during his SPS, *if at all possible.* We realize that the COVID-19 pandemic has thrown a wrench into this kind of activity. We can't advise on the risk to you of taking your puppy to meet other dogs with their people, but we hope you have a safe way to do so. Keep abreast of the advice from the World Health Organization, the Centers for Disease Control, or the government health organization in your country. (WHO and CDC are linked on our resource list at PuppySocialization.com/resources.) We'll discuss this further in the "Dogs from Outside the Household" section below.

It can be unwise for an author to spend a lot of time describing how to do things wrong. That kind of information sticks in people's heads all too well. But we know the information is already out there, being repeated and passed around. Marge sees the fallout every week. You will encounter these myths if you haven't already, so we had to address them. So we ask you to consider those things as the myths they are. We will teach you how to socialize your puppy carefully, properly, and most important, enjoyably. It is great fun to expose one of these little beings to the things that will be in their life. You can watch their understanding, their skill sets, and their whole world grow.

Are Puppies Blank Slates?

They are not. There is so much we can do to affect their outlooks on life when they are 8–12 weeks old (and before that when they are with the breeder). But some things do come preprogrammed. We are learning more and more about behavioral traits that are heritable.

The clearest evidence that behavior is heritable is right in front of us: different breeds of dogs. People have been breeding dogs for specific behaviors for centuries. That's why a border collie is more likely than a beagle to have "herding instinct," and why it would be hard to train a Russell Terrier to be friends with your pet rat. Sure, there are exceptions. Your Portuguese water dog may hate water, or your golden retriever may not retrieve. But the reason we have breeds at all is because dogs were bred for certain behaviors, then immersed from puppyhood in environments where those behaviors were useful or rewarded. Fun fact: working border collies look a lot less alike than the ones from kennel club registries because the working dogs have been bred for a certain type of herding behavior over their appearance.

Why are we talking about dog breeds here? Because in addition to inheriting a propensity for certain tasks, dogs can also inherit behavioral tendencies. For example, tendencies toward fearful and aggressive behaviors are heritable as well (Houpt, 2007). So your fearful puppy may not be the best service dog candidate. There may be limits to what your puppy may be able to happily and effectively do in his lifetime.

So as far as blank slates? Definitely not. Some things come pre-programmed—both behaviors we like and behaviors we don't. And external factors and events before you get your puppy (sometimes even before he is born) influence his behavioral trajectory. No, puppies are definitely not blank slates. At the same time, you can maximize your influence on your puppy's long-term behavior during his sensitive period for socialization.

The Puppy Culture Program

Breeders, shelters, rescue groups, puppy foster families, take note! We are learning more about what puppies need before they leave their dam and littermates. There are some wonderful resources now and one of the best is Jane Killion's innovative program Puppy Culture (Killion, 2014). We believe it should be the standard of care for anyone raising puppies from birth. We include a link to it on PuppySocialization.com/resources.

The Puppy Culture Program explains puppies' physical and behavioral milestones and outlines what breeders and others caring for litters can do to positively influence behavioral development. Attending to puppies' physical needs is not enough. They need exposure to novelty, household noises, challenges, problem solving, and more. The puppies' behavioral development can be enhanced by incorporating some easy-to-follow protocols into their regular care. The reason we are so enthusiastic about the Puppy Culture program is that some behavioral problems can be prevented during the early part of the sensitive period for socialization, before puppies go to their new homes.

This is crucial work. As we said above, when puppies become adolescents and adults, what typically determines whether they stay in their home is their behavior. "Behavioral problems are the number one cause of relinquishment to shelters" (American Veterinary Society of Animal Behavior, 2008a).

The following video shows the puppies in Zip's litter at 7 weeks old. Their breeder provided novelty through a variety of objects. You can see toys of different shapes and textures, objects to navigate through to get them used to entering an enclosed area, and a wobble board to help them get used to unstable surfaces.

The red line with the white ovals is part of the equipment Portuguese water dogs use in water dog competitions. This was a great choice since all these puppies went to homes where the owners would do dog sports. You can also see a grooming table in the background. Zip's breeder did a wonderful job preparing these puppies for life in their new homes.

Video 1.2 The Portuguese water dog puppies explore and play

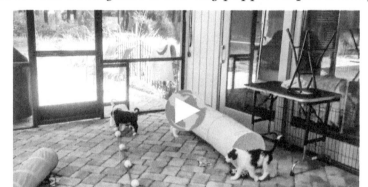

Everyone rearing puppies has something to learn from Puppy Culture.

Chapter 2. Pairing and Priorities

We're both pretty practical people. So we decided to tell you the mechanics of what to do to make your puppy's experiences with the world positive ones before we (briefly) go over the supporting science. This method is so important that we are introducing it as early in the book as we can.

When we talk about "pairing," we mean combining new things and experiences with things the puppy already loves. When this is done correctly, the good feelings the puppy gets from the things he already loves can transfer to the new ones.

New Things Predict Good Stuff

If puppies are such little sponges, can't they just get used to things as they are? Why use food and play?

This is the heart of the book. The goal of socialization isn't merely exposing your puppy to new things. The goal is to help your puppy form positive associations, using food and play, with the things he'll encounter during the course of his lifetime. We want to help you teach your puppy to love the rest of the world as much as he loves food and play.

And here is a preview of how you'll do it (we'll go into more detail later):

1. The puppy sees, hears, or experiences a novel thing.

2. Good things happen for the puppy (usually food, and/or play, and social interaction).

There's more to it; we'll teach you how to read your pup's body language and make decisions based on his responses. But do memorize this now:

New thing first, then food, play, and/or fun interaction.

There will be times when things get chaotic and you accidentally reverse the order. There will be times when everything is happening at once. Don't worry; you won't break your puppy. Simply return to presenting the new thing first as soon as you can.

We want to show you the result of a successful socialization experience using food. Look at this photo.

That face. That's exactly why we pair new experiences with good things, like food and play.

About 40 minutes before this photo was taken, this puppy entered the space for my (Marge's) puppy confidence and socialization class. He was hiding behind his owner, his tail was tucked between his legs, and he moved slowly and cautiously. He would not take offered food. I encouraged his owners to let him observe from a partially barricaded space we created for him in a corner of the room. People and other puppies were not allowed to go into his space.

As he became more comfortable, he started eating delicious food offered by his owners. Then every time a person or puppy passed the space near his area, his owners delivered yummy food.

Soon, he came out on his own to watch the other puppies and the people. His owners offered food as he approached the items set out for the puppies to explore. Pairing his new experiences with memorable food and allowing him to progress at his own pace yielded amazing results in a short amount of time. His transformation during one class seemed nothing short of miraculous. That's the beauty of the sensitive period for socialization and using memorable food.

Let's review two things we touched on already. First, fear is natural and will come bubbling into your puppy's responses to new things, sometimes when you least expect it. Second, lots of the things you want to introduce your puppy to will not be fun or reinforcing. They will be weird to him, sometimes downright scary. Things like the vacuum cleaner, getting nails trimmed, and riding in the car? These are some of the things your pup will likely need more help with.

Pairing Experiences with Food

We need to transform novel, weird, or scary things into good things. And one of the best ways to do this is to make them predict something the puppy loves, like fabulous food. Roast chicken. Roast beef. Homemade tuna treats. Canned dog food. *Memorable food.* If weird or startling things are followed by a bite of something yummy, those things will likely stop being weird or startling. They will become predictors of great stuff. The pleasure of eating good food will actually start to transfer to the weird things, and the weird things will come to evoke good feelings.

So remember: new thing first, then yummy food.

Be ready with really good food. You'll also use his kibble, of course. Your puppy's tummy can't tolerate a constant input of rich food, and it could negatively affect his nutrition. But when your pup is afraid of something, or you know it will seem extra-weird to him, pull out the good stuff.

You might be wondering what that looks like. We'll show you. The puppy in the following video doesn't have a neutral opinion of people. She is worried and unsure about people, but the process is the same. Two people are approaching. We are at a distance where the puppy feels safe and there is an added benefit of a physical barrier, a stream. When the people approach our space bubble, we begin feeding special food (baby food in this case). We *open the bar*—a term coined by the renowned trainer Jean Donaldson. In the presence of the people, the puppy gets special food (the food bar is open). When the people leave our bubble, we *close the bar* (stop feeding).

You can go to the video page to see what that looks like.

Video 2.1 Wilhe gets yummy food as people pass by

You can also use toys and play, and we'll tell you how. But using food is the easiest and most efficient way, and eating is a calming activity. Sometimes you don't want to add in the extra excitement of play.

But Somebody Told Me Not to Use Food

We have seen people claim that using food when socializing puppies is not based on science. This is false. First, the science that supports using food is called classical conditioning, and it has been studied for almost a century now. Classical conditioning is the best way to change an animal's emotional response to something it fears or is cautious about. Second, we know that puppies, even during the sensitive period, can fear novel things (Coppinger & Coppinger, 2002, p. 115). This is true even for normal, well-adjusted puppies. And it's especially true for pups who are genetically fearful or who didn't get the best start in life. Using food in the socialization process is a basic application of strong science.

We have also seen the claim that we shouldn't use food because "mother dogs don't socialize their dogs with cookies." First, this claim ignores the fact that puppies gain comfort from nursing. But most important, socializing their young to the human world is not something any animal does. We won't find examples of it in non-human animals. Socialization is the job of the human!

Now, if handing out food sounds a little robotic to you, don't worry. We still want you to praise, talk to, and touch your puppy (if he likes those things and is not fearful). Social interactions with your puppy are also important and, for many puppies, reinforcing. As a matter of fact, there may be times in your puppy's development that social interactions with you are more important than food. But those times may be hard for you to predict or identify. Because of that, during the SPS, we want you to purposefully use food and play as your primary tools. But you can also layer social interactions on top. "Good boy! You are the best tugger ever! Get it, get it, get it!" That way you gain the benefits of both. Talking to and touching our puppies comes naturally to many of us. Using food and play is something we have to learn. That's where we want your focus to be.

Pairing Experiences with Play

Did you know that puppies who play with their humans are "less likely to show fear during play in an unfamiliar place" (Tóth et al., 2008)?

Play is an important part of my (Marge's) puppy training program for a variety of reasons. First, typical puppies come hardwired to play. They often chase things that scurry away, they wrestle and play with their littermates, and they explore the world with their mouths. If you watch a litter of puppies, you will see them exploring and playing. Play requires cooperation and give-and-take, which are important skills for a dog to learn.

Since playing with a puppy takes longer than giving him a treat and is incompatible with some kinds of exploration, pairing things with play looks a little different from pairing them with food.

What you will notice about my video examples is that the pairing is often with a situation, environment, or experience, rather than an individual object. So the pairing looks more like: Go to a quiet park—play. Go to a part of the park where more is going on—play.

What if we go to a new park and he won't play? Let's say I have cultivated a good relationship with my puppy at home using play, and he never passes up a chance to play. If he can't play in a new location, that's good information. It prompts me to look closer at his body language. Is he displaying the signs of fear and anxiety because he's too worried to play? Or is he distracted because he is confidently exploring the new space and smells? I can use this information and my puppy's body language to help me decide what action to take.

Play can be a good barometer of emotional state for many puppies (Held & Špinka, 2011). Generally speaking, play and fear are incompatible. And the ability to play sometimes disappears before the ability to eat when a puppy is getting worried.

In the following video, you will see a young puppy playing with her owner near a playground (how appropriate!). There are children, other adults, and even another dog nearby. This puppy is readily engaging in play with her owner.

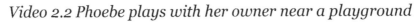

Video 2.2 Phoebe plays with her owner near a playground

Play has even more benefits. We are explaining them in some detail here because using play is less familiar to most people.

Play is a great way to tease up your dog's nervous system to increase his excitement or arousal level. It's important to teach puppies and dogs to recover or calm down after they get excited. The well-known trainer Leslie McDevitt (author of the *Control Unleashed* series of books, linked at PuppySocialization.com/resources) calls this an "off-switch" (McDevitt, 2007, p. 154). If you develop good play skills with your puppy now, it will benefit you later in your dog's life, too.

For example, if you only ever practice obedience skills when your puppy or dog is calm, you're only teaching him to respond to cues when he is calm. But what happens when his favorite person walks through the door, or he sees a squirrel or a deer? His heart rate increases, his respiration increases, his body releases adrenaline. If we never practice cues with him in a heightened physiological state, we haven't taught him to respond in that state. He's not "being stubborn, dominant, blowing us off, or giving us the paw." We simply haven't taught him to respond to cues when he's that excited. So when we are teaching our puppy skills, it's important to incorporate training with various levels of excitement or arousal.

An advantage of using play is that you can control the situation and not go straight into high gear. You can control the puppy's access to the toy and how exciting the toy is.

In the next video, you can see me (Marge) playing with a puppy. I gradually add more movement to the toy to make the game more challenging. Generally, a still toy is less exciting than a moving toy. A slow-moving toy is less exciting than a fast-moving toy. It's generally easier for the pup to resist a toy that's farther away than one that's close. You can't control how the squirrel or deer moves and behaves. But you can control the movement and intensity of the toy. So you can start at a level appropriate for a tiny baby puppy and increase the difficulty as the puppy gains experience and practice.

The puppy in this video is just learning to play with toys with a human. If he looks tentative to you, it's because he is very young and he has never done this before. He's still figuring out the moves. But the activity is already fun enough for him that he will perform a sit to get access to the toy even while in a new and fascinating place.

Video 2.3 Young Maverick plays with a toy in a new place

If your puppy enjoys playing with you, you have a valuable, non-food way to reinforce him. Research suggests that when animals play, it enables the formation of emotionally-based bonds between dog and owner (Bradshaw et al., 2015). The pleasurable emotional experience enhances the human-animal bond and can get associated with you and the socialization situation that you're in.

It's good to intersperse calm activities, such as eating from a food toy or chewing on a safe chew stick, with more intense activity like playing.

Pairing Experiences with Social Interaction

We almost didn't include this section because we don't want you to equate social interaction with the powerful tools of food and play when it comes to socialization. Social interaction means paying attention to, talking to, praising, and petting your puppy. One of the reasons it is weaker for pairing is that you are already doing those things frequently, not just for new events. So it's hard to use social interaction alone to help your puppy form positive associations because it doesn't happen exclusively in the presence of the new things. And most trainers find that the positive effects are weaker than the power of food and play. The occasional puppy might value it more highly in general or during a phase, but those puppies are the exceptions, not the rule.

That said, we want you to use social interactions with your puppy when you use food and play. Also, when it just feels good for both of you. We don't have a video to show you using social interaction alone because Marge has never used social interaction by itself for socialization. But you will see her using it and teaching her clients to use it all the time in videos where food and play are also being used. Social interaction with you is an important part of your puppy's behavioral development. Those interactions grow your relationship with your puppy. But, during this critical period in his development, the sensitive period for socialization, we want you to pair new experiences with food and/or play and not just social interaction.

They fit together nicely, though. All three overlap, and sometimes even happen at the same time! And you'll develop a feel for the best tool for the moment. Food is usually a calming activity, and play is a lot more exciting. There are times when it's best to stick with food and calm social interaction so you don't ramp the puppy up.

Generally speaking, I (Marge) use food combined with social interaction with me initially. I gauge the puppy's interactions with me and his body language and introduce play as soon as I am able.

What Happens If You Expose the Puppy to Things without Using Food or Fun?

I (Eileen) watched this happen and I can tell you.

I live in a suburban area. I've got a big yard that's enclosed with a chain-link fence. The way the houses are laid out, my yard abuts and shares a fence with four other yards. There is also a clear view of the street in front of my house and a cul-de-sac with more houses visible from the back. So, between about eight houses and yards and a couple of streets visible, there's always something going on.

As it happened, the neighbors on either side of my house both got puppies around the same time. One was a Labrador retriever, the other a large mixed breed. The Lab was an "only dog" and the mix had a very nice border collie mix "sister."

When they were baby puppies they were indoors a lot, but as they grew up, they were outdoors much more often.

Because of the layout of the yards in the neighborhood, they could observe a lot. They were regularly exposed to the following:

- At least five other neighbor dogs of different breeds and sizes, ranging from friendly to indifferent
- Adult humans doing human activities
- Children playing noisily on equipment
- People doing yard work
- Roofers
- A lot of construction work
- Cars, cyclists, and delivery trucks
- People walking dogs on leash

So compared to a typical socialization list, that's not a lot, but it should cover some ground, right? Most of these things were in sight but not right up in the dogs' faces, just as we advise you to start when exposing your puppy to new things.

But neither dog ended up comfortable with other dogs, adult humans or children other than their owners, or anything else on the list.

I observed the two dogs as they became more and more fearful, including of me, although I was on the periphery of their lives from the day they arrived at their homes. I made an effort to be non-scary. I used soft body language, never came too close to "their" fences, and spoke to them in a sweet voice when outside. Meanwhile, they looked at me with worry or huffed or barked at me. I also made sure my dogs didn't do anything to intimidate them from our side of the fence. I couldn't ethically give them treats without permission or I would have considered it!

Those two dogs were exposed to quite a few things. Their needs are taken care of and their owners are kind. But both dogs are afraid of almost everything that goes on outside their own yard, even though they were "exposed" to these things from puppyhood on. These exposures were not planned or controlled, and most importantly, they were not paired with the things that make life good for puppies: food and play.

One more mini-heartbreak: of the two pups, the one who lives with the friendly and sociable dog grew up to be more fearful than the other. So there's another idea to put to rest: the theory that if we let our puppy hang out with a well-adjusted dog, it will rub off on him and he will grow up to be calm and sociable. Unfortunately it rarely works that way.

This is only one story. Marge could tell you a thousand. And they all confirm what science tells us: most puppies need controlled, positive exposures to grow up comfortable and happy in their world.

But They Don't Do This in Other Countries and I Never Did This Before and My Dogs Were Fine

Here in the United States, our relationships with our dogs have changed over the past several decades. Dogs went from being primarily outside creatures, often allowed to roam the neighborhood, to members of the family who live inside and become part of our daily lives (Grier, 2010). As rural areas became more suburban, fenced yards became more common. Today a puppy can go home to a neighborhood and see few people outside his immediate family. The way we live now has changed the way we need to raise our puppies.

If you are too young to remember those days and still got a friendly, well-adjusted dog without socializing—congratulations! Through circumstance, chance, or genetics, you were fortunate. We see that sometimes, too. A busy family brings home a puppy and they don't do any active socialization. But because of where they live, the number of family members and their ages, or the puppy's genetics, he turns into a great family pet. On the other hand, another family brings home a puppy, they research and try to do everything right, and they still end up with a fearful or aggressive dog.

We don't have a crystal ball, and neither do you. But we see every day the unfortunate and sometimes tragic results when puppies aren't socialized. Why take the chance? If you're fortunate, your puppy is going to be a part of your family for a long time. Why not take what you learn about how puppies develop behaviorally and stack the deck in your favor? Give your puppy the best chance to live a confident and happy life.

Identify Your Priorities

In order to socialize your puppy properly, you'll need to consider what kinds of environments, things, people, noises, handling, and animals he will likely encounter in his life. Some of these are easy to identify. For instance, if you live in a high-rise apartment, your pup will need to learn about elevators. He likely rode in one with you the day you brought him home. When he's older, he'll likely ride one with you several times a day, and navigate busy hallways, too. But some dogs will never set foot in an elevator. Elevators are still great things to teach all dogs about, since you never know what life will bring. But they aren't as high on the list for someone living in the suburbs or a small town.

Some items require more thought and planning. For example, if your children are grown and live out of town, but you expect visits from your grandchildren in the future, you need to expose your puppy to children now. Waiting until later in the year may be too late. Likewise, do you want your dog to be a therapy dog or maybe compete in agility some day? Guess what? You should start exposing your puppy to those environments now.

After you have figured out your priorities, you can start exposing your puppy to those situations and items in small, fun doses, always making sure the experience is a positive one. Some experiences will delight your puppy. For example, some dogs love water and your pup may be thrilled the first time he gets to visit a stream.

At the same time, many experiences will be neutral or include the possibility of your pup getting worried. That's why you should always have food and toys with you to pair with the experience.

It's not surprising the chocolate Lab puppy in the following photo took to the water immediately!

Pet vs. Sports/Competition Dog: Should My Socialization Goals Be Different?

We just told you to be thinking about your puppy's future so that you can expose him to the things you expect will be part of his life. Does that mean that if your puppy is going to play a sport or have a special task to perform, his basic socialization should be different?

We don't think so. Whether you want to earn an advanced title in a dog sport or have a dog who can welcome visitors to your home and go with your family on hikes, the central goal of socialization should be to produce a confident dog. Confident dogs don't worry if you move a plant in the living room. If they startle or get scared, they bounce back. And because they are confident, they get to do more things, have more adventures, and spend more time with their people.

Having a dog who has been conditioned during their sensitive period of socialization to have positive associations with people and other dogs will only help you with your sport. As your puppy gets older, you can focus your puppy's training and exposure on your specific goals, including attention on you. But during the SPS? Teach your puppy the world is a safe place full of fun things.

If your dog is super friendly to all humans and you take part in a sport that requires focus and attention, you can teach focus and attention as your puppy gets older. Humans will just be another distraction. On the other hand, if you don't adequately socialize your puppy during the SPS, you could end up with a dog who is worried around people and dogs. That would be a problem for you **and** your dog, and not only when he competes. It would be a problem when he walks down the street, when he goes to the vet, or when you have family over for the holidays. It's lots easier for a puppy to learn to ignore humans when appropriate if the puppy isn't afraid of them.

We probably don't compete in the dog sport you are considering or have your breed of dog. But we can still counsel you on what's best for your dog. This is because trainers everywhere see the inadequately socialized dog every single day. Dogs who can't accept a puppy or friendly dog in their home, dogs who bark and lunge at the neighbors when they walk by the yard, and dogs who can't attend even a basic manners class, let alone compete in a dog sport.

If you need more proof, look at the service dog organizations. In almost all cases, the biggest challenge is not the individual service tasks the dog needs to learn. It's being able to perform those tasks in the big, wide human world without fear. It's the "public access" part of the job. That's why many organizations use volunteer puppy raisers—it enables their purpose-bred puppies to get out and see the world from the context of a family home.

Chapter 3. How Puppies Learn and How Socialization Fits In

This is not primarily a puppy training book. We are not going to give instructions on how to teach your puppy to lie down, walk on leash, or even how to go in a crate, although these are all great things for your puppy to learn. But this book *is* about teaching your puppy. You *are* teaching him the most important things of all at his young age: the world is safe and even fun. Believe us, those are big things for your puppy to learn.

We want you to know about the ways animals (and people!) learn that are pertinent to socialization. The first two we'll cover are classical and operant conditioning. Those two types of learning are intertwined; they can't be separated out in the real world. But what we actually *do* when working with our puppies, the actions we take, will often lean toward one or the other. For instance, sometimes we'll recommend automatically feeding your puppy when something like a loud noise happens. Other times, we'll say to feed your puppy when he does something, such as looks at or approaches something. This is because we are seeking to take advantage of one of these two types of learning.

Classical Conditioning

Classical conditioning is how "pairing novel or weird stuff with good stuff" works.

Can you remember a sound from your childhood that came to predict something you liked? I (Eileen) can think of two right away: the sound of popcorn kernels rattling into a measuring cup and the sound of my dad's keys when he came home from work. Neither sound was meaningful to me the day I was born. But as I grew up, since they were always followed by something I liked—eating popcorn or getting to see my dad— the sounds themselves began to elicit the same happy responses.

Dogs make associations just as we do. I bet your puppy already knows or is learning the mealtime-predicting signals in your home. Perhaps the clang of a bowl, or the opening of a bag or can? Does his tail start to wag? Does he drool? The cool thing about the drool response is that the dog's body does it as a reflex. The dog doesn't decide to salivate. It's one of many reflexive responses the dog doesn't control.

The name of this type of learning is *classical conditioning*, sometimes called *Pavlovian conditioning*. You've probably heard of Ivan Pavlov, the physiologist who studied conditioned reflexes. Pavlov discovered that if a previously meaningless sound (or other event) preceded food enough times, dogs started to salivate when they heard the sound, just as they would when presented with food. How strange that is—but also, how lucky for us! It means we can transfer the happy feelings associated with food and play to new people, objects, and locations!

The most important part of this type of learning is that the **thing** (the scientific word is *stimulus*) **predicts** (comes before) the **yummy food or play**. Not just occasionally—reliably. Meet a new person; get some yummy food. To get the positive association, the person or other thing *must* appear before the food. There might be times when you will give your puppy food before the stimulus. For instance, you might keep your puppy busy and distracted with some wonderful food during his vaccinations. You will likely bring the food out before the action starts. That's okay. We call that management. But "food-as-distraction" doesn't help him form a positive association (conditioned response) with the thing/stimulus. To get the conditioned response the thing/stimulus should appear before the food. The timing is important. And we'll keep reminding you so you can keep it straight in your mind. We know that many of you are learning too!

Classical conditioning is such an important tool that you'll probably continue to use it throughout your dog's life. Scary and painful things are going to happen, no matter what we do. This powerful law of learning is one of our most effective tools to minimize their negative effects.

In the adjacent photo, you can see Zip drooling. I (Marge) have a predictable daily sequence at breakfast time. When I finish the sequence, Zip gets to lick my breakfast bowl. My routine is regular and predictable. As the sequence nears completion (my spoon scraping the bottom of the bowl), Zip begins to drool. He doesn't decide, "I hear the spoon clinking, I think I shall drool now." It's not something he can  do voluntarily. Nor did his body have that response to a spoon noise when he was born. His body now does it reflexively, because the noise has predicted food dozens of times. That's a great example of classical conditioning.

If you're thinking, "Hey, Zip is just drooling because he smells food!" good for you for considering that. But Zip, like most dogs who live in a home, smells and sees food being prepared and eaten by humans all the time. He doesn't drool every time. It's the clink of the spoon—the specific clink on the bottom of the almost-empty bowl—that tells his body something tasty is coming his way.

It's important to know that your dog doesn't have to drool for classical conditioning to be happening. All sorts of physical responses and emotions can be conditioned. Dogs who like to go for walks often get excited when you pick up the leash. The leash reliably predicts something the dog enjoys. It also works the other way, too. Dogs can learn that going to the veterinarian predicts something unpleasant. Many dogs pant from stress in the vet's office or even in the car on the way there because they've learned that even the route predicts a vet visit. They made the opposite association to what you want. You want your puppy to learn that the vet is his friend. We'll tell you how to get classical conditioning to work in your favor to teach him that.

Operant Conditioning

We can't talk about raising and teaching your puppy about the world without discussing *operant conditioning*. This is the other main way animals learn, and the well-known psychologist B.F. Skinner discovered and explored it. Pavlov was about building associations; Skinner was about the animal learning through consequences.

Animals learn to make things happen through their behavior. While most of us have to think for a moment about an example of classical conditioning, we are more familiar with operant conditioning. Puppy sits, and you give him a cookie. He is likely to sit more often.

The puppy in this photo is learning to target a hand with his nose, which is another example of operant conditioning.

Marge often reinforces simple behaviors on her socialization outings. Some of her favorites are hand targeting, sit or down, and settling on a mat. You can find more of her favorite exercises in Leslie McDevitt's *Control Unleashed: The Puppy Program* (McDevitt, 2012). We link to all of Leslie McDevitt's books and DVDs at PuppySocialization.com/resources.

A Bit about Positive Reinforcement Training

In real life, the laws of learning—classical and operant conditioning—are almost always intertwined. Your puppy is always learning whether you think you are teaching him or not. He is learning that he can make things happen with his behavior (operant conditioning). And he is learning from the associations that occur in his life (classical conditioning).

Those associations are one of the reasons that veterinarians who specialize in animal behavior issued a position statement against using punishment-based training methods. Their professional organization, the American Veterinary Society of Animal Behavior (AVSAB) has several important and useful position statements. In their position statement about punishment-based training, they recommend that "training should focus on reinforcing desired behaviors, removing the reinforcer for inappropriate behaviors and addressing the emotional state and environmental conditions driving the undesirable behavior" (American Veterinary Society of Animal Behavior, 2007).

For those of us who had a dog more than 20 years ago, there was mostly one way to train a dog back then. We jerked the leash and said "heel" a lot. We were told we had to be a good "pack leader." Guess what? We know better now. Dominance is out. A different AVSAB position statement (linked at PuppySocialization.com/resources) emphasizes that "animal training, behavior prevention strategies, and behavior modification programs should follow the scientifically based guidelines of positive reinforcement, operant conditioning, classical conditioning, desensitization, and counterconditioning" (American Veterinary Society of Animal Behavior, 2008b).

Why reference those position statements in a book about socialization? Because, as we pointed out, classical and operant conditioning are intertwined. When learning is a pleasant and positive experience, those pleasant feelings attach to the behavior, the training environment, and even to you.

For example, teaching your puppy to keep four feet on the floor when greeting people using food as a reward might also help him feel good about meeting new people. Compare that to throwing a can full of pennies near your puppy when he jumps up. That method carries the risk of associating that scary event, and being scared, with meeting new people. He may become more worried about strangers or even aggressive. It may stop your pup from jumping up, because the can of pennies will likely startle or scare him. But you could hurt or traumatize your puppy.

So you have a choice. You can choose methods that create positive associations for your puppy that he will look forward to. Or you can choose methods that could scare him and that he might try to avoid. Whatever you choose, emotions will get built into those behaviors. We do **not** recommend those "old-school" training methods. **We hope and highly recommend that you use positive reinforcement training.**

Speaking of positive reinforcement training, as you watch some of the videos in this book, you'll sometimes hear a noise right before the puppy gets a treat. The noise might sound like a mechanical click, or it might be the word "yep" or "yes." It might even be a mouth click, like you might "cluck" to a horse. Those noises are used in training as *markers*. We call them markers because they're used to mark the exact moment the animal does something we want. At first, those noises mean nothing to the dog, but we give them meaning by having them predict food for the dog (as Pavlov taught us!). The marker tells the animal that the behavior they were doing when they heard the sound is what earned them the reward, or *reinforcement,* that followed. A marker is different from praise, though for some dogs, praise is a reinforcer. Markers, however, tell the animal exactly what behavior earned the reinforcement. It's a subtle but important difference.

Markers enable us to communicate precisely with dogs. I (Marge) have used markers to teach dogs tricks, complicated behaviors, and to stop chasing the squirrel and run to me. Marker training is a powerful tool that can enhance communication with your dog. We encourage you to explore marker training. *Puppy Start Right: Foundation Training for the Companion Dog* is one of our favorite resources for owners (Martin & Martin, 2011). Check out our resource list for some more recommendations on the topic (PuppySocialization.com/resources).

Habituation vs. Sensitization

There are two other learning processes that are important to know about. They are habituation and sensitization (Domjan, 2018, p. 31–36). These are ways animals, including humans, can respond to their environment over time.

Habituation happens frequently without our even recognizing it. We (humans and animals) can get used to something initially strange or even startling when it appears repeatedly. For instance, humans can get used to the feeling of comfortable shoes on our feet, glasses on our faces, even contact lenses in our eyes. These are very novel sensations at first, but we soon forget about them. An example with puppies is wearing collars. Initially, a puppy might scratch at a collar, but he will usually get used to it very quickly.

Sensitization is the opposite of habituation. Sometimes our response to an experience or sensation grows over time instead of diminishing. For example, I (Marge) have been wearing contact lenses for years. I don't even feel them in my eyes any more. My husband tried wearing contacts and the sensation of *something in his eye* grew each time he put them in his eyes. He did not get used to them. He wore them less and less. He became sensitized to them and stopped wearing them. An example with puppies is a pup who is startled or scared by his first holiday decoration. Whether it was a larger than life snowman or a "ghost" in a tree, he could end up more afraid of holiday inflatables. I hear this all the time from clients.

So how do we know ahead of time whether a dog will habituate to something or get sensitized? We can't know for sure, but most scientists say that the more intense, scary, or painful an experience is, the more likely the animal is to sensitize to it. Dr. Pamela Reid says, "In general, really intense stimuli often lead to sensitization while weak stimuli usually habituate" (Reid, 1996, p. 137).

A good rule of thumb is to never assume that a puppy will get used to something that startles or scares him. He might. But puppies don't usually habituate to things that scare them. What often happens is the opposite. The puppy's worried reaction becomes stronger. He becomes sensitized.

Even though puppies are more open to new experiences during the SPS, they are not immune to sensitization. They can get scared of something and that fear can grow. We can't ever depend on habituation happening. We need to help along the process of getting used to the weird human world. We do that using food, play, and social interaction.

Bank Accounts

So, as tempting as it is, we can't put your puppy in bubble wrap and protect him from bad experiences. They are going to happen. What can we do?

Many trainers use the concept of creating "bank accounts" of positive experiences for puppies. You can pair good stuff with the things he's likely to encounter during the course of his lifetime: children, other dogs, new places—all the things we mentioned earlier. Every time your puppy has a good experience with a thing, person, or place, you are making a deposit in the bank account for that experience. Then, when real life happens and something startles your puppy and takes a big withdrawal from the account, there are still some good feelings left from prior deposits. And you can work to top it up again!

Continuing with the bank account analogy, through good genetics and/or prenatal and early experiences, some puppies are born with behavioral "silver spoons" in their mouths. They are resilient, curious, and hit all the behavioral milestones. They are born with deposits already in their bank accounts. Other puppies, through genetics or a stressed prenatal experience, are born with negative balances in their accounts.

And that is exactly why you should use food and play when you expose your puppy to the world. Nothing puts a deposit in a bank account like something the puppy needs and loves. Make no mistake; there will be withdrawals. So make sure you keep each experience bank account topped up well enough to withstand the occasional withdrawal.

Review: Use Food and Play to Form Positive Associations

We can't assume puppies will habituate to everything in our human world. They will get worried by some people, things, or environments. By the time the average person notices that their puppy is worried, he has likely been uncomfortable for a little while. That's why we want you to use food and play *automatically* when your puppy encounters something new. The goal isn't for the puppy to have a neutral experience with new things, people, places, and experiences. The goal is to have **good experiences and positive associations.**

Chapter 4. Understanding Dog Body Language

Reading dog body language is essential to socialization. If you can't recognize the subtle signs of fear and anxiety in your pup, you don't have the necessary skills to socialize him. We know it's tempting to skip this part and get to what feels like the good stuff: what to do and how to do it. Please don't. The ability to read basic dog body language is the difference between success and failure. You can't effectively socialize your puppy to the world if you can't tell what he's saying with his body language.

What you learn here will help with more than socialization, too. An understanding of dog body language will help with training (for both puppies and adult dogs) and can strengthen your bond with your puppy. Knowing what he is "saying" to you with his behavior will make all the difference—for both of you.

Before we review the body language of fear in dogs, we'll help you establish a baseline for non-fearful dogs. So let's look at the set of behaviors that generally combine to show us a relaxed and happy dog. If "relaxed and happy" sounds a little ambiguous right now, don't worry. We'll show you a variety of different puppies and dogs and show you exactly what we mean.

Body Language Examples of a "Relaxed and Happy" Dog

It may seem unnecessary to review the body language of relaxed dogs. Can't we all tell when dogs are relaxed? We just look at them and get a feeling that the dog is relaxed. But our feelings can be wrong, and they don't teach us what to watch for. We're going to teach you observable clues or cues of relaxed dogs. There are commonalities in the body language of relaxed dogs. And once we recognize them, they provide a nice

contrast with the body language clues and cues that tell us a dog might be worried or anxious.

First, let's take a look at a photo. This is the goal of socialization: to create a "relaxed and happy" puppy when he encounters something new. Experiencing new things is an adventure and you want your puppy to enjoy the adventure.

As you introduce your puppy to things that make up life with humans, remember the expression on this puppy's face. (And don't worry—lots more examples are coming.)

Let's look at more physical characteristics of what we're calling "relaxed and happy." The puppy in the next video is relaxed and comfortable in this environment. She is leaning in for petting.

Video 4.1 Relaxed Cede leans into a hand for petting

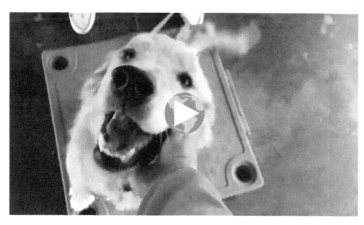

All the following photos show several aspects of relaxed and happy behavior. But we have chosen photos that show at least one behavior very clearly, and labeled them accordingly. In real life when you try to assess if your puppy is relaxed and happy, you won't just look at one body part. You'll look at his whole body.

- **The mouth is open and relaxed, like a smile. You'll notice this commonality through almost all the photos. It's a dead giveaway.**

- **Muscles look loose and soft.**

- **Eyes are soft and squinty (not staring).**

- **Pupils are appropriate for the setting (not large in moderate to bright light).**

- **Weight is distributed equally.**

- **Ears are held in a relaxed manner appropriate for the breed (Example 1 of 4).**

- **Ears are held in a relaxed manner appropriate for the breed (Example 2 of 4).**

- **Ears are held in a relaxed manner appropriate for the breed (Example 3 of 4).**

- **Ears are held in a relaxed manner appropriate for the breed (Example 4 of 4).**

- **Hip is rolled and hind legs are relaxed when lying down.**

- **If the dog is panting, the panting is relaxed and easy.**

Video 4.2 Bria's panting is relaxed

- **Tail is relaxed and wagging or swishing in a relaxed way (Example 1 of 2).**

- **Tail is relaxed and wagging or swishing in a relaxed way (Example 2 of 2).**

Video 4.3 This dog's tail is relaxed and swishing

Look at the Whole Dog

We've pointed out that you need to look at the whole dog. Look at everything and take in the different signs. Occasionally a dog's behavior will be slightly atypical, but with a careful look you can generally determine whether he is relaxed.

Here is a good example of that. We've said that a relaxed, open mouth is generally a sign of a relaxed and happy dog. The dog in the following video has her mouth closed. But you'll notice some of the other things we've shown so far that indicate she is happy and comfortable: soft, squinty eyes, ears held in a relaxed manner, and soft loose muscles. Look how relaxed the front of her mouth is. As she leans quietly into being petted, you can see this dog is relaxed and happy, too.

Video 4.4 Summer is enjoying being petted—she is relaxed but with a closed mouth

Relaxed and Happy: Mini-Quiz and Homework

How many of the happy and relaxed characteristics we have discussed do you see in the following photo?

Now take a look at some of the photos you've taken of your current or past dogs. Can you find photos showing these characteristics of relaxed and happy dogs?

Body Language of Fear and Anxiety in Dogs

Now we are going to discuss the signs of fear, stress, or anxiety and show you examples.

We've learned a lot about canine communication and body language in the past few decades. I (Marge) first came across information about canine body language in Dr. Patricia McConnell's books, *The Other End of the Leash* and *For the Love of a Dog*. Later, I found the late Dr. Sophia Yin's work through her blog posts and her book, *"Low Stress Handling, Restraint and Behavior Modification of Dogs & Cats."* We're going to use Dr. Yin's free, downloadable handout, "Body Language of Fear in Dogs," as a guide in this section of our body language discussion. The handout is linked at PuppySocialization.com/resources.

As we start to look at the signs of fear and anxiety in dogs, it's important to know that context matters. Yawning is often listed as a signal of stress, and we include it in the list below. But a dog who yawns and acts sleepy while lying in a dog bed at home isn't necessarily stressed. On the other hand, if your dog is at the vet's office for his yearly checkup and yawns, it could be a sign of stress. The context is different.

Here's an example. I used to teach at a dog training club. I was there one night when a colleague was teaching a beginner dog training class. It was the first night of class and, boy, could you tell it was a beginner class! All the dogs were doing some type of "naughty" (undesirable) behavior, such as pulling on the leash, barking, sniffing, or jumping up. But one dog was lying quietly at his owner's feet, with his head on his paws. It looked like he was sleeping. Someone elbowed me and said "Look at that dog.

He's so well behaved!" I told her that the dog was scared. That he was doing the doggie equivalent of putting his fingers in his ears and saying, "Lalalala." He was actually extremely stressed, what trainers refer to as "shut down"—too scared to move.

Most of the other dogs were doing typical things for untrained dogs in a new location. They were smelling doggie smells, sniffing for varmints, barking at the other dogs, and interacting with the environment. For the dog lying down, given his age and level of training, taking a nap did not fit in that context. Lying down and taking a nap your first time in an exciting environment is like going to sleep the first time you go to a professional football game. It means something's wrong. That dog was overwhelmed.

As you learn to read your dog, it also helps to know that signs of fear and anxiety often occur in clusters, not isolation. You will often see several signs occur in a short period and the dog will look stiff and tense, not soft and wiggly.

Dogs communicate all the time with their body language. It's happening whether we notice it or not. For instance, dogs almost never bite "out of the blue," although we hear the phrase frequently. Dogs give lots and lots of subtle warnings using their body language. For your dog's well-being and your own, you need to learn about these warnings and other signs that your dog is distressed. We recommend that you print out Dr. Yin's handout so you have an easy reference.

Dr. Yin lists the following behaviors as indicators of fear and anxiety:
- cowering
- licking lips (when there is no food nearby)
- panting (when not hot or thirsty)
- brows furrowed (with ears to the side or back)
- moving in slow motion (walking slow on the floor)
- acting sleepy or yawning (when they shouldn't be sleepy or tired)
- hypervigilant (looking in many directions)
- suddenly won't eat (but was hungry a moment earlier)
- moving away (including trying to escape)
- pacing

Other things to look for are hiding, freezing, trembling, front paw lifted, tucked tail (take breed into account), weight shift away, approach-avoidance behavior (see below), whites of the eyes showing (also called "whale eye" or "half-moon eye"), and enlarged pupils when it's not dark.

Approach–Avoidance and Displacement Behaviors

These are two kinds of stress-related behaviors that can be harder for people to detect. Puppies who are anxious or unsure often demonstrate a behavior that is referred to in psychology as approach-avoidance conflict. The American Psychological Association defines it like this:

> *"A situation involving a single goal or option that has both desirable and undesirable aspects or consequences. The closer an individual comes to the goal, the greater the anxiety, but withdrawal from the goal then increases the desire"* *(American Psychology Association, 2021).*

I (Marge) see this conflicted behavior a lot and can tell you about a classic example of it. I met a puppy for the first time and she came running up to me with a wagging tail, wiggly body language, but a closed mouth. And when she reached me she retreated a few feet. Then she approached again and retreated again. She did the same thing with the novel objects I brought with me. Puppies, especially young puppies, rush up to greet people or investigate things because they are curious. But if they are a bit nervous, unsure, or even had a bad experience in the past, they will often then avoid the person or object. So just because your puppy rushes up to someone or something, it doesn't mean he is comfortable. Keep watching.

Other behaviors that can be signs of mild stress are called "displacement behaviors." Displacement behaviors happen when animals are conflicted. As the well-known author and behavior consultant Steven Lindsay writes:

> *"Displacement activity occurs when some course of action is thwarted (frustration) or when two opposing motivation tendencies are elicited at the same time (conflict). Under the influence of frustration or conflict a substitute behavior may be emitted. ...the displacement activity may be activated whenever the animal is confronted with a difficult or insoluble conflict, providing a mechanism for safely killing time until a more adequate response can be found to resolve the situation"* *(Lindsay, 2001, p. 135).*

Displacement behaviors in dogs include abruptly sniffing the ground, scratching, excessive licking, and even drinking water. All can be signs of anxiety, although you have to look at context when assessing these behaviors. If your pup is sniffing while trailing a chipmunk, he is probably not doing it out of anxiety! (We'll show you a video later of sniffing as a displacement behavior.) It's the context that matters. Not every dog who scratches or sniffs is stressed. But, if you raise your voice to your dog, escalating a cue or command, and he suddenly begins to scratch, he might very well be stressed. Look at what other clues you can get from the dog's body and the context.

Heeding Our Dogs' Subtle Signals

Here's a way to think about the more subtle signals. Picture two adults talking over a fence between their yards and one of the adults has a little boy along. The little boy notices a fuzzy, crawly thing and it scares him. He's not sure what it is. He keeps his eyes glued to it and he tugs on the back of the adult's shirt. The adult figures the kid is safe because the yard is fenced in, and he keeps talking. The little boy has his eyes glued to the scary thing and the scary thing starts moving. So he tugs on the back of the adult's shirt and the adult keeps talking. Now the thing is moving toward them— what *is* that fuzzy crawly thing? The little boy tugs harder on the adult's shirt. The adult keeps talking and absently puts a hand down and comforts the boy. The caterpillar is still crawling along and finally the little boy tugs really hard and yells, "Hey, hey, HEY!" The adult finally turns around and says, "What?"

These more subtle signs by your dog are his way of tugging on the back of your shirt. When something in the environment scares him, he can't use words to say, "Excuse me, owner. I've never seen a little girl with a shark fin on her bike helmet and I'm a little nervous." He's going to use the only language he has—his body language. Lip lick. Yawn. Brow furrowed. And we need to heed these less obvious signs. That's because some dogs—if they tug on the back of our shirt and they tug on the back of our shirt and we don't listen—decide to take further action. It's as if they're saying, "Fine—if you won't protect me, I'll protect myself." And they start growling, barking, lunging, air snapping, or biting. Your job as an owner is to pay attention to the "tug on your shirt" and intervene. Increase distance from the thing that's worrying your dog. Listen when he tugs so he doesn't have to escalate his behavior and "yell" to get your attention.

Body Language Examples of a Fearful or Anxious Dog

To supplement the fear signal drawings on Dr. Yin's poster (have you printed it out yet?), here are some photos and videos of dogs showing various signs of fear and anxiety. Some of the dogs in the following examples have been diagnosed with clinical levels of fear or anxiety and have been treated by a veterinarian or veterinary behaviorist.

• Cowering (Example 1 of 2)

When dogs cower, you often see that the outline of the top of their body (called the "topline") is rounded or curved. This is often referred to as "roached." The tail will be low or tucked and often the dog will bend his leg joints, lowering his body closer to the floor, almost like he's trying to make himself smaller.

The puppy in the next photo has his tail tucked so tight that the tip of it is right against his belly. His topline is rounded, his joints are bent so that he's almost in a crouch, and his ears are to the side.

• **Cowering (Example 2 of 2)**

This dog's back is roached as well, and she is leaning away, so much so that her paw lifted off the ground.

• Licking lips/tongue flick

A tongue flick is one of the most common stress signals in dogs, and sometimes it happens so fast it is hard to see. This tongue movement is different from that of a dog who is anticipating a tasty treat. Context will help, and over time you'll learn to tell the difference between the lip lick your dog does when he's anxious and the "Yum!" lip lick.

The puppy in this photo is nervous because she's in a new environment. Notice how her tongue has flicked up over her nose.

• Panting when not hot or thirsty (Example 1 of 2)

This pant is short, shallow, and rapid. Often the tongue is straight out—not draped over the teeth and lips. It is not the same as when your dog pants after you play a vigorous round of fetch, with his mouth open and the tongue hanging out to the side. Also, the open-mouthed expression in this photo is different from the relaxed mouths in the previous section. Note the very tight muscles around the corner of the mouth (called the commissure).

This dog is nervous in this new environment. During this session, she displayed many of the signs of fear and anxiety we are reviewing.

- **Panting when not hot or thirsty (Example 2 of 2)**

The next video shows a dog at doggie daycare. (What dog wouldn't like being at daycare, right? You'd be surprised, but that topic is outside the scope of this book.) The kennel environment is stressful for this dog. Notice her panting; it is rapid and shallow. How is this panting different from relaxed and happy panting? (Note before you play the video: there is a dog barking.) You can compare this video to Video 4.2 above that shows Marge's dog Bria panting in a relaxed way.

Video 4.5 This dog's panting shows stress

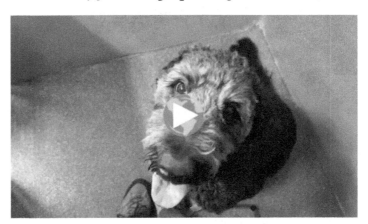

Remember how we showed you that a dog with a closed mouth can still be relaxed? The two dogs in the previous photo and video demonstrate the parallel situation, that a dog can have his mouth open but *not* be happy or relaxed.

- **Brow furrowed—ears to the side or back (Example 1 of 2)**

When dogs worry, they often get wrinkles on their forehead, between their eyes. Their ears are often to the side or all the way back.

At first glance, this puppy in the following photo looks like she has a cute and wrinkly face. But look more closely. She's certainly cute, but there is a lot of tension in her face causing those wrinkles. Her skin is taut over her face, her mouth is shut, and her ears are back. Even though she is comfortable with the people accompanying her, she is worried about being in a new environment.

- **Brow furrowed—ears to the side or back (Example 2 of 2)**

While this puppy does not show wrinkles in her brow, her ears are down and back to the extreme. She is worried.

• Moving in slow motion/sniffing

The classic examples of slow motion movement showing stress can be seen in the dog "shaming" videos on YouTube. They make our hearts hurt. The intentions are to be humorous—supposedly the dog is acting "guilty" because he knows he did something wrong. But in those videos, the dogs move slowly because they are worried or afraid of their owners at that moment. Dogs are experts at reading our tone of voice, facial expressions, and body language. They know what a certain combination predicts. For instance, "Bad things happen when my owner raises her voice and points."

The following video doesn't quite show the typical slow motion move, but you can see that the dog is tentative and slow. This is Eileen's adult dog, Zani, who is moving slowly because she is very anxious. The repeated ground sniffing is a displacement behavior, another sign of stress for her in this context. One way to tell is that she doesn't seem interested in what she is sniffing. It's almost like she's killing time.

Video 4.6 Zani is moving very slowly and showing other signs of stress

• Acting sleepy or yawning (Example 1 of 2)

The anxiety-related "acting sleepy" behavior is when a dog seems to rest or act sleepy in a very out-of-context place or situation. Like at a children's birthday party or sporting event. Of course, young puppies tire easily and they will often fall asleep if they are stationary too long. Notice how your puppy behaves when he first arrives somewhere.

In the following video, Eileen's young dog Clara isn't sleepy. She's at the veterinary clinic, yet she yawns. And look at the rest of her behavior. She is agitated and can't settle down. She pants, and as in our previous examples, her panting is not relaxed. So is the yawn a sign of settling down for a nice rest? Definitely not.

Video 4.7 Clara yawns from stress at the vet's office

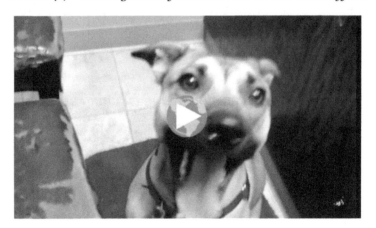

- **Acting sleepy or yawning (Example 2 of 2)**

The puppy in the next photo is not yawning because she is sleepy. She is stressed in a new location.

- **Hypervigilant**

Dogs that are hypervigilant are constantly on the alert for things that scare them. They often have trouble settling and staying in one place. They look all around for scary (to them) things to appear.

The dog in the following video is hypervigilant and keeps looking in all directions. This dog is worried about storms and it is raining. In the video, you can see her turn toward noises, even "typical" household noises.

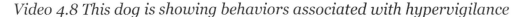

Video 4.8 This dog is showing behaviors associated with hypervigilance

• **Suddenly won't eat and moving away**

One moment the dog or puppy is eating and the next moment they are not.

The puppy in the next video shows examples of not eating and moving away. She is afraid of the chickens that are outside. I (Marge) am sitting in a corner. Notice how the puppy is trying to escape and moving backward. She eats a little bit of food, but leaves other food. The puppy put herself into the tight space between the couch and coffee table. I wouldn't ask a nervous puppy to get into such a confined space.

Video 4.9 This scared puppy backs away and leaves tasty food

• Pacing

When I (Marge) think of pacing, I picture a person walking around a room, wringing their hands. Dogs have a version of that.

The dog in the following video is not interacting with the environment at all. She paces continually and also refuses one of her favorite foods.

Video 4.10 This dog is pacing

• Weight shift away (Example 1 of 2)

This means the dog or puppy is leaning away from the scary thing.

The puppy in this photo is worried about her owner moving something.

• **Weight shift away (Example 2 of 2)**

This puppy walks by these water bottles in the house daily. When she sees them out of context (outside), she is worried.

• Approach-Avoidance

We presented the definition of approach-avoidance above. Dogs displaying approach-avoidance appear conflicted. They often approach something or someone willingly or with some encouragement. At the same time, the closer they get to the scary (to them) thing, the more worried they become. They often display their worry by immediately moving away and retreating, or approaching cautiously, with their weight shifted back.

The puppy in the following video is conflicted. She wants the yummy food treats and is still very curious, yet she is frightened of the new objects. I (Marge) use that food toy to assess puppies when I meet them initially.

Video 4.11 This puppy repeatedly approaches, then leaves food toys

- **"Whale eye" or "half-moon eye" (Example 1 of 2)**

"Whale eye" or "half-moon eye" means that the whites of the dog's eyes are showing in a way that is not typical for the dog. This usually occurs in combination with other body signals, and the dog looks scared. Sometimes the whites of the eyes look like a sliver of the moon, as shown in the next photo.

This puppy is very worried in a new environment.

- **"Whale eye" or "half-moon eye" (Example 2 of 2)**

This puppy is worried about someone to her left. Everything about her body language tells us she is uncomfortable.

• **Mouth is closed or clamped shut**

One of the easiest things to notice is whether the dog's mouth is open or closed. That does not mean every dog who has his mouth closed is stressed. It's a starting point. You still need to consider the situation and note what other cues and clues the dog might be giving you with his body language. As you look back through this section, notice how many photos show the dogs' mouths open and how many show the mouths closed.

The dog in the next photo just moved to a new household. Her closed mouth and other body language cues tell us she is not yet "relaxed and happy" in her new environment.

- **Dilated pupils**

  A dog's pupils normally grow larger in response to low light, but if the pupils are dilated in normal or bright light, the dog is probably afraid.

  Note how the dog's pupils in the following photo nearly fill the dog's eye, even in bright light.

Look at the Whole Dog (One More Time)

We're going to conclude this section with two photos. Both show dogs exhibiting many signs of fear or anxiety. The first photo shows Eileen's dog Zani in a moment of deep distress. She demonstrates several of the signs we've discussed. Her tail is tucked, her back is curved (roached), her ears are back, her facial skin is pulled taut, her mouth is pinched shut, and her weight is shifted back.

This second photo is one you've seen before. This is Eileen's dog Clara. You watched a video of her vet visit in the stress yawning example.

We showed you this photo of Clara in chapter 1 and asked you to take a look and think about whether it showed a relaxed or stressed dog. We feel confident that you know the answer now. Take a look at her brow, cheek muscles, tongue, and pupils. She's very stressed. You'll learn more about Clara later in this book.

My Dog Is "Fine"

We hear this a lot. A dog looks a little nervous in a situation and the owner says, "He's fine." Those owners are at a disadvantage because they haven't learned what you have. Many people think if a dog isn't barking, growling, lunging, trembling, or hiding that he is "fine." We cannot emphasize enough that the more subtle signs of fear and anxiety in dogs are every bit as important as the more overt signs. When you introduce your puppy to the world, you don't want a puppy who is merely "fine." You want a dog who looks "relaxed and happy."

For example, if you take your dog downtown to see the sights and he looks like the dog in the following photo, he is not "fine," even though he's sitting quietly.

What about the dog in the following photo? Can you tell why she is not "fine"? You can see the breakdown in our body language quiz at the end of the chapter.

Sometimes our dogs' communication stays in the background for us until they are "yelling" (sending strong signals) for help. We notice when they are barking. We notice when they are trembling. We notice the overt signals of fear and anxiety. But we don't always notice when they are sending subtle signals, when they are "whispering."

It can be difficult to see what a puppy is "saying" when we are overwhelmed by how cute he is or when we have tasks to get done. Sometimes it helps to compare how the dog looks when he is in a new situation and how he looks when he is "relaxed and happy."

Body Language Comparisons

In this section, we feature paired photos of the same dog showing the dog when they are comfortable and relaxed versus scared and worried. Look at the photos below and list how many differences you see, and how well they align with the body language elements we have listed above.

Back to Context

As we mentioned at the beginning of this section, context matters. A dog can be very focused on training or a game and not be relaxed at all. But neither is he fearful. The important thing is whether the dog's behavior is appropriate for the circumstances. Look at the context. And always keep in mind that still images can be deceiving. The

dog in this photo looks like she's displaying "whale eye," but the photo was taken just before she turned her head to look at something. But while we are at it, can you see any other clues that this dog is not scared?

That's Eileen's dog Zani again. Even though her mouth is closed, her facial muscles aren't tight, and her brow is smooth. But it would be easier to tell with a comparison. We have one. You can go back to page 85 to see her in deep distress.

It's also good to keep the above photo in mind whenever you look at any still image of a dog. Photos show us the dog's position for a fraction of a second. We need to look at still photos to learn, but a still can be deceiving. Also, dogs don't helpfully freeze in position in real life when you want to examine their body language. That's why we are showing you lots of videos!

Since we've shown you Zani in distress, Zani moving slowly because of anxiety, and Zani doing "whale eye" even though she wasn't upset, we wanted to show you what happy Zani looked like. Can you see how relaxed and happy she is in, of all places, an obedience club? She loved other dogs and all people. What signs of relaxation and happiness do you see?

How to Greet a Dog: Greeting and Lifting Up Puppies

How to greet a dog might sound like a no-brainer. But you'd be surprised how many normal things people do make dogs uncomfortable. The more you learn about dog body language, the more you will start to see their discomfort. In the meantime, here are some pointers about how to greet and pick up dogs in ways that are more comfortable for them.

"How to Greet a Dog and What to Avoid" is another great handout from the late Dr. Sophia Yin. We link to it at PuppySocialization.com/resources. I (Marge) include this handout in my sessions with my training clients. It's not only got fantastic graphics on how to greet a dog and what to avoid, but it also communicates the larger message: our body language impacts how our dogs respond to us. I'm not going to review the handout in detail, but I do want to point out two things.

First, be sure to follow Dr. Yin's instructions: approach sideways and look using your peripheral vision. Dogs do not like uninvited direct eye contact. From the time human babies are born, we start making eye contact to bond and show affection (picture a mother feeding a newborn baby). It is not the same for dogs (picture a mother dog feeding a litter of puppies). Direct eye contact between dogs who don't know each other is usually a sign of trouble brewing. So is a head-on approach. By turning sideways, with your shoulder facing the dog, you appear more friendly and less confrontational.

Second, respect the puppy's space. Think of dogs as having space bubbles around them, as the graphics show in Dr. Yin's handout. Some dogs have small space bubbles, and some dogs have great big space bubbles and they need more personal space. When your body presses into, or invades, a dog's space bubble, you end up pushing the dog away from you. Have you seen the "close-talker" on Seinfeld? It's the same thing. (We link to the video at PuppySocialization.com/resources.)

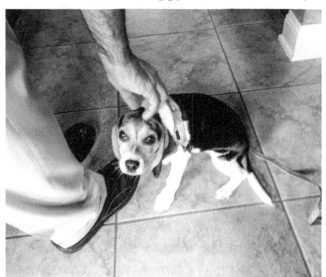

What do you think this young puppy feels about the person leaning over into her space? Is she "relaxed and happy?" Look where the man's foot is, and imagine how much of his body is looming over the puppy as he bends so far over. She has shrunk down away from his touch, her back is roached, ears are flat, and tail is tucked.

In the following pair of photos, the same puppy is greeted two different ways on his first day home. Which human is greeting in a way most puppies would prefer?

If you guessed the second photo, you are right! The man with the big hat is leaning over the puppy's space bubble. Most puppies don't like that. The man in the second photo got on the puppy's level and let the puppy come into his space.

We want to acknowledge that plenty of people lean over puppies because other physical options might be painful or unavailable to them. Many people are uncomfortable squatting or physically can't do so. Also, sitting on the ground is not an option for everybody. When possible, sitting in a chair to greet the puppy is a good alternative to leaning over into their space with one's whole body.

Lifting Up Puppies

When most owners lift up their puppies, they lean right over into the puppy's space bubble. The puppy often starts shrinking away or avoids being picked up. Professional trainers usually know what has likely been happening when a client says of a young puppy, "I can't catch him!"

What's a better way to pick up your puppy? Point your shoulder to your puppy. Collapse at the knees (instead of bending at the waist) and invite your puppy into your lap. Secure him with your hands and then stand up. That's much more comfortable for your puppy than leaning over top of him (into his space bubble), grabbing him around the ribs, and suspending him dangling in mid-air as you pull him close to your body.

Again, we know that not everyone can comfortably and safely perform this maneuver. If you can't, watch your puppy's body language and figure out a way to pick him up that is safe for you and the least unpleasant for him. Perhaps he can get on a short stool or even a step. Marge teaches older puppies to hop up on a low ottoman or bench. And whether your puppy is on the ground or an ottoman or a bench, don't forget to pair lifting with a tasty treat.

Marge made the following video on how to lift a puppy for her private training clients.

Video 4.12 Marge demonstrates picking up a puppy

Owners tell Marge all the time that their dog or puppy seems to like some people but not others. A big part of how a puppy reacts to people can be how the people interact with him. Are they putting their face in the puppy's face? Or are they allowing the puppy to approach at his own speed? Think of your puppy's behavior as information. It is information about what he likes (is he rushing toward you?) and what he doesn't like (or is he moving away?). Listen to what your puppy is saying with his behavior.

Another Thing to Avoid During Greetings: Don't Rub That Belly

Speaking of space bubbles, one thing that really puts pressure on a puppy's space bubble is leaning over them during greetings. Some puppies, in response to this intrusion, will roll on their backs and offer their belly. They are typically not asking for a belly rub. Puppies or dogs who offer their belly, often with a low or tucked tail, are showing "withdrawal from interaction," showing "concern" or "true fear" (Overall, 2013, p. 135).

The scientific term for this behavior is inguinal display. Sarah Kalnajs, in her DVD The Language of Dogs, descriptively calls it a "tap out," like the wrestling term (Kalnajs, 2006). It's the puppy's way of saying, "I give!" Or "I'm not a threat—here's my belly." We want confident postures when puppies greet, not puppies on their backs.

In a typical scenario, a puppy first rushes in to greet someone. The person leans over the puppy, puts two hands on the puppy in a familiar way, and the puppy rolls over. His tail, though wagging, is usually doing so while plastered to his belly. The owner and the greeter think the puppy is asking for a belly rub because he eagerly ran up to the new person, then flipped over. But that's not what is going on at all.

Here's the thing: puppies are curious. They might be excited about meeting new people, even rush up because people often have treats for puppies. Then the human does the looming thing and reaches with both hands into the puppy's space bubble. Remember the space bubble around the dog on Dr. Yin's handout, "How to Greet a Dog and What to Avoid"? Check for the link on PuppySocialization.com/resources. Spatial pressure alone can cause some puppies to "tap out."

Some puppies do sometimes like belly rubs, but you need to look at the context. If you're petting your puppy on the couch in a slow, relaxed way and he's floppy and loose, stretches out and rolls on his back, he might want a belly rub. Do puppies ask for belly rubs during greetings? Almost never. That is almost always a response to physical or social pressure. It is the puppy's way of saying, "Please stop what you're doing." When an owner tells me (Marge) their puppy loves belly rubs that is usually a tip that an owner is misreading his dog's body language.

Watch the following video closely. This puppy rushed up to greet a visitor to her home, then flipped over. That's where the video starts. She's met this person before and always "asks for a belly rub." But she really isn't doing that. Notice where the puppy's tail is. Notice how very stiff and tense she is. Dogs who are relaxed and wanting to be petted are generally loose and floppy. See if you can detect her trembling. She also pushes against the visitor's hand with her feet, mouths her hands, and yawns.

Video 4.13 This puppy shows many signs that she doesn't want a belly rub

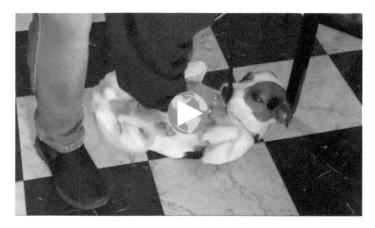

Take Our Quiz

Test your dog body language knowledge with our quiz.

Video 4.14 Take the quiz

How did you do? If you are new to studying body language, congratulations for taking the time to learn how dogs communicate. Your skills will serve you well beyond puppyhood through the life of your dog. If you're a savvy dog owner, with lots of experience, we hope you enjoyed seeing different dogs showing these familiar signals. It's an ongoing learning experience for all of us, and we never get to "perfect." At the same time, we can aim to be good students because there is always something to learn.

Chapter 5. Socialization!
Starting at Home

Word is finally starting to get out that puppies need to be exposed at a very young age to the world they will be living in. They need to be socialized. But socialization is rarely explained correctly, so people who do try it often don't have the information and skills to succeed. That's why we wrote this book. We want you to succeed. We want your puppy to be comfortable in the world and for the two of you to develop a joyous relationship.

So now you're ready to get started on your puppy's socialization. You've reviewed the chapter on dog body language and watched the videos. You've watched the dog body language quiz video on page 101. And if you have kids, you have shown them, too. You've applied what you've learned and have been observing your own pup and other dogs you see. You feel like you have a pretty good idea of when your pup is happy and comfortable.

We talked earlier about socialization checklists. Make a list, use an existing checklist, or use a combination. Dr. Sophia Yin's Puppy Socialization Checklist is very thorough. You can access the link for the free download on PuppySocialization.com/resources. It prints up nicely on standard size paper. An important component of Dr. Yin's checklist is the option to rate each interaction from 1–5. We highly recommend that you do so. As you go along, observe your puppy's body language and match your observation with the rating scale. It will help keep you focused on your puppy's body language and will help you learn to recognize when your puppy needs help. It also keeps you out of the "just check it off" mentality.

It's important to know that all puppies, even the most confident ones, have days they are less confident. This might happen because the puppy encounters something extra worrisome, like a girl with a backpack or a boy carrying a tuba. It is easy to miss those times if the puppy is not cowering and shaking. But if you're using Dr. Yin's checklist and rating the interactions regularly, you are more likely to notice the more subtle signs of fear and anxiety we discussed earlier.

And along those lines, it is not a bad thing when you notice when something worries your puppy and you note it on the checklist. It is a good thing! You can't help your puppy feel better about things that worry him if you haven't identified them. So please don't inflate your puppy's rating or feel like something is wrong with him if his scores aren't all 4's and 5's. Some lower scores mean that you have excellent observation skills and are doing your best to help your puppy be resilient and confident.

Also keep in mind that checklists are guidelines, not rules. **Socialization is not merely exposing your puppy** to the things on the checklist. It's the process of creating **positive associations** with those things.

We've divided socialization types into the following categories:

- Environments/places
- Objects
- People (including children)
- Sounds
- Handling
- Other dogs
- Other animals
- Time alone

Many of these can be found or experienced both at home and out in the world. We will start at home, because that's probably where you will start, and you can exert more control over the environment there to keep from overwhelming your baby puppy.

Creating Great Associations at Home

You might not think of socialization as happening at home, but when you consider it, there are tons of things to introduce puppies to. Your home is an entirely new environment for your puppy. You can take advantage of the seven days at home while you're waiting for the puppy's first set of vaccines to kick in. Some of these things will be introduced as part of your everyday life. Some you can introduce intentionally.

Let's think through our categories (environments/places, objects, people, sounds, handling, other dogs, other animals, and time alone) and consider some examples. Because we are talking about the home environment, any list will be heavy on the objects and sounds categories but will include some of the other categories too. Here are some examples:

- Doors of all sorts and sizes
- Brooms and mops
- Vacuum cleaners
- Hair dryers
- Unstable surfaces that move underneath the puppy's feet, like skateboards or wobble boards
- Leashes
- Collars or harnesses
- Family members wearing unusual clothing (see "What If the Pandemic Prevents You from Introducing Many New People" below)
- Human visitors
- Other dogs
- Sudden household noises like accidentally dropping a pan
- Loud music/sounds from TVs and devices
- Mail, deliveries, truck noises, someone coming on porch
- Garbage pickup
- Drones or other remote controlled devices
- Minor accidents (we'll explain this below in "Unexpected Sounds and Events")
- Bathing, grooming, and husbandry
- Other animals
- Being alone

That's only a sampling of the things in your home that may be new to your puppy. We want to get you accustomed to considering your home from your pup's point of view: it's a new world. And then we want you thinking about your job: creating good associations.

Remember How to Create Good Associations?

Use food. And you will use both *good* food and *great* food. The "good" food will usually be your puppy's regular dry food, referred to as kibble. You can use his meals, or part of his meals, for training. It will be part of his daily ration, not an *addition* to his daily ration.

Part of the deal when you got your pup was feeding him. Why give all that valuable food away for free? Puppies have to learn about and get used to so much: handling, strangers, the vet, nail trims, kids—you get the idea. Use some of his calories to help him feel good about the world. For most puppies, kibble will work fine for teaching them behaviors like sit and for introducing them to non-scary things at home. But when things are harder or scary, you'll need great food.

Marge's go-to items for great food are real meat (cooked) and cheese cut into tiny pieces. Eileen creates baked treats with leftover meat that has been chopped fine in a food processor. We both also use canned dog food and spray cheese in moderation. Some commercial treats are okay, but be aware that quality, packaged treats generally cost a lot more per pound than an expensive cut of beef. So why not buy some decent beef on sale, roast it, and cut it into tiny, puppy-sized pieces? You can keep a handful of pieces in the fridge and freeze the rest.

As you introduce these new foods to your puppy, make sure to use only a little bit. Puppies have sensitive tummies. And naturally, when you add special food to your puppy's diet, you have to cut back on calories elsewhere in his daily ration. Making these adjustments will become second nature.

You can also use some of this special food to help introduce your puppy to the joys of stuffed food toys. Kongs are popular, and the Toppl by Westpaw is also a great introductory food toy for puppies—the food is more readily accessible, even by very young pups. You can soak some of your puppy's kibble, mix it with some pieces of the really good stuff, and put the mixture in the toy. As your pup learns how to get food out of the toy, you can make it more challenging by packing the food in more tightly, and finally by freezing it.

As usual, it's better to go slow than rush things. Once your puppy can manipulate the toy to get the food out, it is a great tool for socialization (think car rides, veterinary visits, grooming) as well as enrichment. You can vary the value of the food for different situations, again, keeping in mind your puppy's daily nutritional needs. But definitely mix in some tasty, stinky stuff the first few times while he gets the hang of it, to keep his motivation high.

We use the word "treat" to mean food of any value. We are not saying you should feed your puppy "treats" in addition to his regular food. Any training food should be counted as part of his normal, daily caloric intake and will frequently be his regular food.

Building Your Skills

Now that we've given you an overview, it's time to practice your skills. Wait, what? That's right. Introducing your puppy to the world requires specific skills.

One skill is observation. If you haven't studied canine body language before buying this book, you won't become a skilled observer all at once. But don't worry. The more you observe your puppy's body language, the better you'll get at it. After all, he's talking with his body language all the time. You'll definitely want to get better at learning to "listen" to what he's saying.

There are also mechanical skills such as treat delivery (where, when, and how you present the treats) and timing.

We won't go into detail on all the mechanics of treat delivery, because, after all, this is a book on puppy socialization, not dog training. But one thing to know about very young puppies is that they sometimes need help finding treats when they roll away. They don't understand the physics of treat movement yet! More important, they may be startled by your tossing motion. So while away from your puppy, practice "bowling" a treat (gently rolling it on the floor underhand) onto the floor near an object.

To practice treat "bowling," you'll need some of your puppy's kibble. Put it in your pocket or have some nearby. Stand about one or two feet away from an object in your home and get a piece of food. Keeping your arm close to the side of your body, roll the treat with a gentle "bowling" motion of your arm and hand. Squat down a little if you are tall, but don't lean over. Remember our discussion about puppies' space bubbles? "Bowl" treats a few times, varying between aiming straight in front of the object or a little to the left or right.

It's good to practice with different types of food. Moist meat and cheese often don't roll far; they might even just go "splat" and stay where you toss them. Kibble and dehydrated treats roll, but sometimes erratically. The surface of the floor makes a big difference, too. But keep your eyes on the prize. Get a feel for how treats move. Right now, more important than where the treat ends up is the smoothness of your bowling gesture.

When you begin rolling treats to your puppy, if he doesn't see a treat you tossed, tap gently on the floor next to the treat with your fingers. You'll likely only have to do this a couple of times. Pups learn quickly how treats move!

If you pursue training your puppy (this is not a training book, remember?) you'll learn that treat delivery and placement are important factors and require a lot of skill. But right now what's important is that your pup finds the treats and isn't scared by your movement.

What You'll Need to Get Started

To start the socialization process, you'll need:
- A hungry (not starving) puppy. You should make whatever adjustments to your puppy's meals necessary to achieve that result.
- Your puppy's kibble. You can jazz up the kibble a little. If your puppy is not that interested in his regular food, try putting some of it in a baggie with a hot dog or other aromatic food. The kibble will absorb some of the smell and taste of the other food and this will usually make it more appetizing.
- Great treats, as described on page 106.
- Toys.
- Household objects (see list below in "Video Examples of Building Positive Associations with Household Objects").

Remember how we talked about classical conditioning and making good associations? Here is where you put it to use. First, make sure you have some of your puppy's kibble and some special treats accessible to you at all times. *All* times. We're not kidding. Yes, you are going to be that person who carries treats around in your pocket. Because sometimes life happens. Not all your socialization activities will be planned and set up. Marge jokes that when she has a puppy in the house she always has kibble in one pocket and a toy in another, even when she's in her bathrobe. Hey, you never know if your neighbor will start up his motorcycle on your puppy's 6 a.m. potty outing (this actually happened to Marge) or the garbage truck is going to come by.

You can keep food in your pockets. You can keep lidded jars of treats or kibble (out of puppy's reach) strategically around the house. One of Marge's clients coined the term "kibble vases." You can have kibble in some and higher value treats in others. Some dried and freeze-dried treats that are high value for most pups can be safely left at room temperature for a few days. Two great places for kibble vases are near your puppy's confinement area and near the front door, because you will frequently want to teach and reinforce good behavior in those areas.

What You'll Do: Introducing Your Puppy to Novelty at Home

The first day you brought your puppy home already had plenty of newness for him. So guess what—you've already started socializing! Remember—everything is novel to a puppy. You've probably used things like the dishwasher, hair dryer, and television. Maybe stairs are new to your pup, or sliding glass doors. You have an ottoman or a houseplant, different substrates to walk on, things on wheels, curtains, and so on. You can start exposing your puppy to novelty in a more deliberate way the day after you bring him home.

Your puppy may have gotten a head start on novelty if your breeder, shelter, foster family, or rescue started your puppy on Puppy Culture (linked at PuppySocialization.com/resources). On the other hand, puppies who were raised in isolation (like in a barn, garage, or kennel) will have a harder time learning that new things are fun.

The most important part of introducing novelty is what we call "puppy's choice." We allow the puppy to investigate the novel object at his own pace. We don't try to get him to move closer to something he is afraid of. We use our observation skills and we give him all the time he needs.

We'll first describe the basic process of introducing novelty, then show several video examples.

Introduction to Novel Objects: The Steps

This is a procedure you will follow, with some variation, many times with your puppy. You'll use it at home and in new environments; you'll use it with objects, people, and sounds. Your puppy will be learning about the world, and you will be learning the mechanics of using food and how to observe and manage your puppy. For your puppy's first introduction to novelty, choose something that you think will be neutral and non-threatening, a simple object you would bet $100 won't scare your puppy at all. Look for something solid that stays put and doesn't tower over him and large enough that he can't swallow or damage it. A full, non-opened can of soup is a good starter. Starting with something easy will give you some breathing room as you practice observing your puppy and getting food to him.

1. Get that sturdy, non-scary item and put it on the floor.
2. Be ready to give your puppy food any time he interacts with this novel object. If he's there with you he'll likely head toward it right away.
3. You want to reinforce (that is, give him food for) any behaviors associated with confidence: investigating novel but non-scary objects. "Investigating" includes looking at it, leaning toward it, approaching it, sniffing it, and touching it.
4. Give your puppy a treat after each of these behaviors. Be generous! You don't have to use a clicker or marker, but you can if you want.

Remember to always start in your puppy's comfort zone. Don't start with something you think your puppy might find scary. If your puppy shows a little fear or startles, that's okay. In their sensitive period for socialization, they bounce back pretty quickly. And they are more likely to approach things even if they are a little scary. Remember to reinforce interest, even if it's only a glance. We want the puppy to form positive associations with new things. If your puppy stays scared, read our section on abnormal fear in the SPS. You may need expert help for your puppy.

After you introduce things that are not likely to scare your puppy, you can start introducing more unusual things the puppy might not have experienced before. Things like a metal cookie sheet to walk on, a low, plastic tub or cardboard box containing some plastic water bottles, or a hairdryer. Start with less intense objects and work your way up to more intense objects. Be ready to raise the value of the treats for more challenging objects and experiences. Use the special stuff!

Video Examples of Building Positive Associations with Household Objects

My (Marge's) puppy Zip had a wonderful start at his breeder's home. He and his littermates experienced a lot of new things every day. However, I knew he still had much more to learn about the world. When Zip was little, I set up "confidence courses" in the house, yard, and the driveway. As Zip approached and experienced something new to him, he got lots of yummy food. I used his meals, with a few special treats mixed in, as he encountered these new things. We also played near and sometimes on the novel objects.

Here are some of the things Zip experienced:
• Garbage cans
• Blow dryers

- Walking on different surfaces, like cookie sheets, plastic tarps, and bubble wrap
- Stepping into things that touched his legs (a tub filled with empty plastic water bottles and milk jugs)
- Unstable surfaces (a skateboard and balance equipment)
- Objects with odd shapes and movement (an umbrella)
- Things to walk through (a cardboard box)
- Decorations (a large stuffed bunny)
- A person using crutches
- Noise-making toys (a child's "popper" vacuum)
- Going into a dark space and having something touch him (a cardboard box with a towel hanging from it)
- A baby carrier
- A balloon

You can see in the next video how I incorporate fun and play into object interactions and confidence courses for Zip. There's plenty of food, too. And so much is going on. You can see him climb on, step on, and interact with many of the items listed above. These were all things I had around the house. You don't have to spend a lot of money. Just be creative. Remember: almost all the stuff in your house will be new to your puppy.

Video 5.1 Zip has fun interacting with new items

The next video shows very young Zip enjoying some chewables on a wobble board when he was still living with his breeder.

Video 5.2 Zip has a chewy on the wobble board

Zip experienced all those things in familiar environments. Now, Zip enjoys novel things! When we travel, hardly anything startles or scares him. When he encounters something new, his first response is to investigate. He has received a lot of reinforcement for interacting with novelty. After experiencing different surfaces and

things that wobble, it's no surprise that Zip is comfortable on an unstable surface like an upside down kiddie pool, as you can see in the adjacent photo.

In the following video, I use some common household objects with a client's puppy: a plastic tub full of rinsed milk jugs and water bottles, a cookie sheet, and a skateboard. We put a non-skid bath mat in the tub to help the puppy feel more secure. And before the puppy stepped on the skateboard, the owner held it, so it didn't move suddenly and startle the puppy. With these three items, the puppy can step into and onto things that move or make noise and experience things touching her legs as she moves. All these experiences are paired with food.

Look for the moments in the video when the puppy is slightly wary of the movement of the skateboard. At 1:08 she slows down, pauses, and braces herself a little as she approaches, and at 1:13 she takes a couple of steps back. Each time she recovers so quickly it's over in the blink of an eye. The next time the skateboard moves she doesn't flinch, and shortly after, she puts her front paws on it. That's the beauty of the SPS.

Video 5.3 The border terrier pup interacts with things that move and provide new sensations

In the next video you will see how two different puppies the same age respond when introduced to a novel object, a food toy that moves.

It may not be immediately obvious to you that the Dalmatian puppy is worried about the moving food toy in the video; she is active and exploratory and wags her tail a lot of the time.

But she demonstrates approach-avoidance behavior with regard to the food toy. She is worried about it, and doesn't bounce back quickly like the border terrier puppy did in the previous video after being started by the skateboard. (That puppy was in its SPS; this puppy is not.) Take a good look at her responses, watching as many times as you need to to see how she repeatedly approaches and backs away from or avoids the food toy. Then you can compare her responses to those of the Lab puppy who is also in the video.

Video 5.4 Two puppies have very different responses to novelty

In the previous videos, we are showing you not only how to introduce your puppy to novelty with simple objects, but also what kind of behavior to look for. If your puppy is tentative and anxious about new objects and doesn't bounce back and recover quickly, you may need to hire a trainer or behavior consultant.

We hope seeing some examples conveyed how simple yet how very important this work is.

Guidelines for Introductions to Novelty

Here are a few guidelines to follow as you introduce your puppy to new things:

- **Allow your puppy to approach things at his own pace.** Resist the urge to lure your puppy toward things he is frightened of. There is too much risk involved. He wants the yummy food, but he might still be afraid of the object (or person). It's not a race. Remember: if you reinforce your puppy for interest in the object, even just a glance, you'll likely get more interest. That's how reinforcement works. Not only that, but he'll also be forming positive associations with the object. Using food or a toy is okay to move your puppy around *when he is not frightened.* That's an important distinction. Don't be one of those people who puts treats all over scary objects.

- **Introduce novelty daily.** Make investigating some new object you have in your home a regular activity—a fun game the two of you play together. Then, when you're in public and you encounter something new, like a Halloween decoration or a playground swing, he already knows what to do. You can start playing the fun "investigate something new" game. You want your puppy to think that whenever you bring him to something new in the world, it's something he can play training games with. Joy and curiosity are better reactions than being worried.

- **As the puppy gains confidence, you can increase the intensity of the items.** A hierarchy of things from less intense to more intense will vary by puppy, because they are individuals. But generally speaking, stationary things like garbage cans are less intense than moving things like luggage being rolled by. Quiet things like plastic bins are less intense than noisy things like some children's toys. Predictable things like a cookie sheet are less intense than things that change unexpectedly like an umbrella. A sample hierarchy of intensity at home might look like this (and remember, you should adjust this for your own puppy):
  - a cookie sheet
  - a small waste can
  - a skateboard turned upside down
  - a quiet children's toy
  - a plastic container (on a non-slip surface) large enough for the puppy to put two feet into
  - an unstable surface like a balance disc
  - a musical or noise-making children's toy
  - a pair of crutches
  - an umbrella

Beyond the Basics: Introducing the Vacuum

A rolling vacuum that is running is pretty high intensity for a lot of puppies and dogs. Consider all the ways it's strange. A vacuum cleaner is an odd shape, it moves, and it makes weird, loud noises. So many challenges all at once! Let's say you know your puppy is likely to be afraid of the vacuum cleaner based on his reactions to other things. How can we reduce the intensity of this high-intensity object?

We suggest you split up the sight and sound of the vacuum so you can proceed gradually. Start with the vacuum turned off and unplugged so you can't accidentally turn it on. Put the vacuum on the floor. If it's an upright vacuum, lay it down in a way that it won't roll or wobble. Remember to reinforce your puppy for even looking at the strange object sprawled out on the floor. Make sure to deliver the treat to him where he is standing. Don't use the treat as a lure to get him closer to the vacuum. If anything, deliver or toss the treat a little bit away from the vacuum. Doing so will give you more information about how your puppy is feeling about it. What does your puppy's body language look like after he gets the treat? Does he approach the vacuum again? Is he happily investigating the strange object or is his weight shifted back and low? Don't proceed to the next step until your puppy does not look worried and shows relaxed and happy body language around it.

The next step, when your puppy is ready, could be to stand the vacuum upright or partially upright by leaning the handle against something. Make sure to hold onto it if your puppy ventures close so it doesn't fall and scare him. Toss your puppy a piece of food when you stand the vacuum up. The vacuum just shape-shifted! Again, reinforce your puppy for showing any interest in the vacuum.

Steps for gradually increasing the intensity might look something like this. These steps can take place in different sessions or across different days—there's no hurry.

- Roll the vacuum without turning it on; give pup a treat. Repeat this several times.
- Gradually roll it more and turn it. Give pup a treat for each movement.
- Put it in a separate room from your puppy, then start and stop the motor quickly (you might need a helper); give pup a treat.
- Start the motor and leave it on a little longer, give pup several treats while the motor is on, then turn the motor off and stop the treats.
- Move the vacuum to a closer room and repeat.
- Work up to the vacuum being in the same room and repeat.
- Finally, start to move the vacuum around. Move it first with the motor off again, since the combination of sound and movement is a big deal for many dogs. Toss treats *away* from the vacuum as you do so. Then do the same with the motor on.

Follow these steps a few times before you do any real vacuuming, and you'll likely get a pup who says, "Oh, the vacuum. Cool."

I (Eileen) conditioned three dogs to respond to the vacuum cleaner as the best thing ever. If you don't want your dogs to be this close when you vacuum, you can teach them to go to a mat or somewhere else after they are comfortable with the vacuum. Personally, I am delighted my dogs formed positive associations with the vacuum and choose to spend time with me when I use it.

Video 5.5 Eileen's dogs flock happily toward the vacuum

A Word about Leashes

You'll see images and movies throughout this book of puppies dragging leashes as they play or interact with things. Leashes are a great way to keep your puppy safe and prevent him from rehearsing behaviors you don't like, both inside and outside the home. And they are something that your puppy needs to learn about. The sooner you introduce your puppy to a leash, the easier it will be for him (Scott & Fuller, 2012, Ch. 5 The Critical Period, "Boundaries of the Critical Period"). You'll need him to be comfortable with it before you go on any outings.

You can start by attaching an inexpensive, lightweight leash to his collar or harness and letting him drag it in the house. Directly supervise him whenever he is dragging a leash so the leash doesn't get tangled or caught on something. And we suggest an *inexpensive* leash because every typical puppy is going to chew his leash at some point. Your puppy will learn to follow you by your body language and voice (if you need help with this, consult a professional trainer). By the time you pick up the leash to start actual leash training, the first step will be done for you: he'll be used to having a leash attached to his gear.

People

Introducing puppies to new people is easy and hard at the same time. It's easy because almost everyone enjoys interacting with puppies. Puppies are like magnets to humans. Who can resist that puppy smell and those big eyes?

But introducing your puppy to people is hard because it seems like everyone who wants to meet your puppy is intent on interacting with him on their terms, not your puppy's terms. "Oh, look. He's scared. Isn't that sweet? He just needs some love!" said as the stranger scoops up your terrified puppy and puts their face in his face (puppies hate that!).

Our guidelines below will walk you through some ways to head off that kind of interaction.

You can start the process of introductions while you are still working with your puppy at home. You have more control there than with total strangers out in the world. If you are able, and can do so safely during the COVID-19 pandemic, pick some friends to come visit who are low key and will be happy to follow your instructions. You can meet in your yard if necessary, once your puppy is comfortable and familiar with it.

The adjacent photo shows a puppy checking out a new person and her mask. Can you tell this puppy is a little cautious? This was followed with a bite of extra tasty food and the pup warmed up very quickly! Also, check out that nice, low-key human body language! The new person did a great job: she sat on the ground, turned her upper body away from the puppy, and didn't look directly at her.

Use the same process with people that you have with objects. Teach your puppy that **new people predict food automatically.** That means that you have to be ready to feed your puppy. See an especially tall guy walk by? Puppy gets food. Meet a new person? Puppy gets food.

Don't wait until your puppy seems scared. Make the new person or thing a predictor of super yummy food.

Puppies Meeting People: Some Guidelines

Here are some guidelines for helping puppies interact with new people. These guidelines will help you become an advocate for your puppy and smooth over possible social awkwardness when necessary. Most of them apply "on the road" as well as at home, and you can refer to this section again when you start taking your puppy out. Remember to follow safety precautions to prevent transmission of COVID-19.

1. Introduce a variety of people. People come in lots of different shapes, sizes, and skin tones. They have different types of clothing, hats, gaits, silhouettes (carrying things, sitting, standing, lying down), languages, gestures, and more. It would be impossible to expose our puppies to every type of human available. Nor is it necessary. But it is important, as we expose our puppy to different humans, to include variety.

2. Set up your environment and your puppy for success. Decide ahead of time where you will ask an indoor visitor to sit. For outdoor visits, provide seating for your guest and yourself. Give your puppy a chance to eliminate shortly before your visitor will arrive, and make sure you have some treats and a few of your puppy's favorite toys.

3. Ask visitors to help you before they visit your home. Prepare your visitor ahead of time so they know what to do to help keep your puppy comfortable. Tell them to let the puppy approach at his own pace and get down to the puppy's level, if possible. Ask them to follow your directions and to be patient if the puppy doesn't want to interact with them.

4. Pair the arrival of the new person with food. Visitor comes in the door; you give the puppy a treat. Visitor walks across the room to sit down; you give the puppy a treat. After the visitor sits down and your puppy is calm, let the puppy approach the visitor if he wants. Use treats, distance, and/or play to create good associations if the pup is at all worried. Heck, use treats anyway. Remember, we don't want neutral associations. We want to build positive associations. Let the pup go do something else or leave if he wants.

If you meet someone in your yard, with or without social distancing, the process looks the same. If the puppy can see the car drive up, give him a treat. When the car door opens and a person pops out (think how unusual this appears to a pup seeing it for the first time!), give him a treat. Give a series of treats as the person approaches and sits in the chair you provided. But again, let your puppy retreat if he wants. He may or may not want to approach the visitor.

After your pup has a short opportunity for interaction and if your visitor will stay for a while, give your puppy something to do in his exercise pen, crate, or other safe area he's used to. Give him a food puzzle or a safe chew toy to keep him busy for a bit. This will let you visit with your guest without having to manage your puppy. It will also help teach your puppy to settle when visitors come to your home and let him practice a mini-separation. We describe this in more detail below in the section "Spending Time Alone." Make sure his physical needs (elimination, food, water, exercise) are met before you put him in his pen or crate.

Note that this is not the time to train door manners. (Although, interestingly, this process we are describing will make it easier.)

5. Help your puppy by using the Three-Second Rule. The Three-Second Rule is a way to keep greetings short and successful. It works great when you allow your puppy to interact with strangers and people you are less familiar with. You can also use it with people in your home.

Here's how I (Marge) do it at home. As described above, I explain ahead of time to any potential visitor a little bit about what I'll be doing and that it's the puppy's choice to interact. When we are set up, I tell my visitor that I'll be calling my puppy back regularly during the interaction, and ask them to stop interacting with him when I do that. Then, if my puppy wishes to greet the person, I allow him to greet for three seconds. I count to myself using "one banana, two banana, three banana" while one of us feeds him treats (more on who feeds on page 124. Then I call the puppy back to me using a positive interrupter ("Pup, pup") and give him a treat.

I observe my friend and the puppy's body language the whole time. If the puppy looks comfortable with the interaction and my friend's behavior didn't scare the puppy, I release the puppy to interact with them again. If my puppy is not comfortable, I move him into another room or behind a barrier and give him something nice to chew if I plan to visit with my friend for a while.

The benefits of this process are threefold:

- The puppy learns the behavior of leaving an interaction with a person. We are adding a behavior to the palette of options available to him.
- Calling my puppy back interrupts the greeter's actions. People do unpredictable things to puppies. They get too familiar; they may suddenly pick the puppy up; they might even flick the puppy's nose! (Unbelievably, I saw that happen once!) This is more likely to happen on the road with people you don't know, but again—people can be unpredictable!
- My puppy gets interrupted before he gets so aroused he can't successfully return to me and practice leaving an interaction. The more engaged a young puppy becomes with something or someone, the harder it will be to call him away. Puppies amp up quickly.

6. Puppies should be on the floor or ground when meeting new people.
That way, you'll be able to tell much better whether your puppy wants to greet the stranger. Remember, your puppy's primary ways to communicate with you are through movement and body language. If he's on the ground, you can tell whether he's moving toward the new person, moving away from the new person, or trying to hide behind you. You can check to see whether his weight is shifted forward or backward, whether his tail is tucked or neutral, and whether he is wiggling or stiff. It is also much easier to pair meeting a new human with food (to feed your puppy) if you're not holding said wiggly puppy in your arms. Even small breed puppies should be on the ground, not trapped in someone's arms, when meeting new people.

If you are social distancing outdoors, you can use a 6-foot or longer leash to help measure distance. I (Marge) like using a leash longer than 6 feet. I use a 10-foot leash to give everyone a little extra space and because I don't want the puppy pulling to the end of the leash.

7. No belly rubs during greetings. We mentioned this before, but we know how pervasive this practice is and how hard it is for humans to stop. So we are going over it one more time. Many of us enjoy a good neck rub. But we don't want one from the stranger at the farmer's market. When your baby puppy rolls over when a new person leans over to pet him, it is likely not because he wants a belly rub. Most puppies, and dogs for that matter, do not run up to strangers, get in a vulnerable position and say "hey, do you mind scratching me here?" It is often a response to too much social pressure (pressure to interact).

Humans interpret an exposed puppy belly as exactly the opposite of what it really means. Sometimes puppies go belly up for their owners, too, for some of the same reasons. Maybe the humans are doing something the puppy doesn't particularly like or enjoy yet, like putting him in his crate. Or maybe their greetings are too exuberant or go on too long (there are lots of reasons to keep them low-key).

During greetings with owners or new people, the looming, petting person often overstays their welcome. This can happen within only a few seconds. (Remember the Three-Second Rule!) The person doesn't know to stop to see if the puppy wants more or is done. The puppy doesn't have a big enough palette of behaviors to know he can leave. Or maybe he can't because he is restrained by the petting hands. So he uses the only language he has, his body language. He exposes his belly and asks the human to stop.

Wait! You mean my puppy might not love everything I do to him to show affection for him? We hate to be the ones to tell you, but yep. Not only is that possible—it's probable. Remember the neck rub. We don't usually want one when we are trying to identify the mystery noise in the attic or when we are right on the verge of finding the mistake in a spreadsheet. Unfortunately, we tend to do the equivalent to dogs and puppies all the time, especially if we don't understand dog body language. Also, some people don't like back massages at all, and some dogs are simply not into being petted. So when your puppy is telling you something you are doing to him makes him nervous, don't take it personally. Change your behavior if at all possible.

8. Keep your puppy's mouth busy. Keeping a puppy's mouth busy is also important, especially when your puppy is meeting children. In the next photo, taken before the COVID-19 pandemic, the owner of this cute puppy keeps his mouth busy and helps him keep four paws on the ground. The pup is also forming positive associations with petting from a stranger. That's a big payoff for a few pieces of food.

What about "Glee Pee"?

Marge here. People often assume that puppies pee from happy excitement. This is what I'm referring to as "glee pee." But I see very few cases of *true* excitement urination. What I see most often are puppies who are nervous from too much pressure. The pressure comes from over-exuberant greetings by people, which often put social and spatial pressure on puppies. The "glee pee" response can be similar to the "tap-out" belly exposure we discussed earlier and can happen as the puppy approaches the person or during the greeting itself.

Dr. Karen Overall reports that when a dog goes belly up and urinates, the dog is showing "profound fear" (Overall, 2013, p. 135).

Puppies who supposedly "glee pee" during greetings with their family and friends almost never eliminate during greetings with me. This is because I know not to put pressure on the puppy. I let him approach at his own pace. I use body language that is not intrusive to dogs and puppies (see Dr. Yin's chart on how to greet a dog linked on PuppySocialization.com/resources). I make sure not to lean over him.

I see this kind of peeing a lot in families with children. The children rush in after being at school or outside and run excitedly into the puppy's space. They lean over the puppy or scoop him up, and they make a big fuss. Even if the children wait quietly and the puppy rushes up to them, they often *then* lean directly over the puppy and into his space bubble. And then the pee happens. Most typical puppies want to greet their people, even when there is pressure, so they rush up. But then the humans tend to add to the pressure by doing exactly the wrong thing.

Do yourself and your puppy a favor: keep greetings low-key when you come home. How you greet your puppy each day is how you are teaching him to greet other people. Keep that in mind as you create those rituals. Also, make sure to consult with your veterinarian about your puppy's elimination habits. Too frequent urination, or a sudden change in urination habits, can signal a medical condition. Your veterinarian should be your first resource for your puppy's physical and behavioral health.

Who Feeds the Puppy During Greetings?

Where does the food come from during greetings? Different experts have different opinions (surprise!). The eminent trainer Jean Donaldson recommends having the strangers feed the treats (Donaldson, 1996, p. 63). The late Dr. Sophia Yin, in her book *Perfect Puppy in 7 Days: How to Start Your Puppy Off Right,* recommends *you* feed your puppy almost continuously to "keep her from jumping on kids and guests, since these people will not know how to avoid reinforcing jumping behavior" (Yin, 2011, Section 6.2). There are two take-home messages here. First, two great, world-renowned behavior experts (and many others) agree that you should be using food to help your puppy form positive associations with strangers and new things. Second, you should put some thought into where the food comes from.

Eileen here. I've made the exact mistake Dr. Yin warns about. I've let others feed my dog and they had a great time feeding and laughing while my dog scrambled and jumped all over them. This was an adult dog but she learned the opposite of the polite greeting behavior I intended! I've also made both good and bad decisions about letting a new person feed my formerly feral pup, Clara.

We can offer some guidelines on how to make the decision (and you might make different decisions in different situations). Here are some things to consider. We've already shared the critical importance of reading your puppy's body language, so you will know if your puppy appears worried at all. If he's worried, we want the food (lots of it, and memorable, too) to come from you. If you think the person might not follow instructions—and believe us when we say you'll start developing a sense of this—the food comes from you.

There are situations in which it could work well for the other person to give your puppy the food. A good setup in which you can have another person feed is if:
1. The person is known to you but a stranger to the puppy;
2. The person will follow your instructions; and
3. Your puppy has normal interest and curiosity about people.

Another good time to have the new person feed the puppy is when there is a physical barrier like an exercise pen or baby gate between them.

You will encounter arguments against letting puppies have happy interactions with strangers. Some people, including trainers, want their dogs to be neutral to all strangers and completely focused on their owner or trainer. But we respectfully submit that not gaining positive associations with people is risky for many breeds and individual dogs. Focus on the trainer or owner can be addressed after the SPS, as the puppy matures; it can be trained. Helping a fearful dog after the SPS has closed is much harder. The benefits of building a general positive association to people will pay off in so many ways. It's almost impossible to overdo it.

Let's get back to what our experts, Donaldson and Yin, told us. Both of them say to go way overboard exposing your puppy to the human world during the sensitive period for socialization. Is it a lot of work (and food)? You bet it is! But at this period in your puppy's life, the time investment is much smaller than if you put it off until later. Do yourself and your puppy a favor—make the time now. Your puppy will thank you.

What If the Pandemic Prevents You from Introducing Many New People?

We know that there may be limits on the exposures to strangers you arrange for your puppy. There is something you can do at home, though: play dress-up. No, not the puppy. You and any other humans in the household. Take Eileen's word on this one. She scared her year-old feral puppy when she wore a loud tropical print shirt one day, deviating from her consistent attire of solid-colored tops. Her puppy panicked, thinking a stranger was in the house.

Hopefully, your puppy will be nearer the center of the bell curve than Clara, but they will benefit from seeing a variety of attire, even if familiar people are the ones doing the modeling. Go through your closets, look in the attic, and gradually expose your puppy to out-of-the-ordinary clothing. If you get your puppy in the spring or summer, make sure to drag out your winter clothing. Be sure your puppy sees you wearing warm hats, ski masks, heavy boots, scarves, and your overcoat.

Do you have wigs, a fake beard, big hats? Dressy shoes that sound different when you walk in them? Did your college-age child leave a sports uniform, even a helmet? Don't forget sunglasses, which seem to look very weird to dogs. (And you'll have your treats ready, right?)

This is not a substitute for meeting new people, but it can certainly help for your puppy to see the variety of human attire. So here's an idea: do this even if your puppy *does* get to meet a lot of new people.

We've introduced the concept of starting with the novel thing farther away, such as with the vacuum cleaner, to make it less likely to frighten your puppy. Be aware that if you wear items that change your silhouette, such as big hats or bulky jackets, the difference might be startling to your puppy even at a distance. So be extra careful with this process. Strange as it may seem, changes in your appearance that you take for granted can be startling or even distressing to your puppy.

As with all things, let your puppy be your guide. If your puppy shows you with his body language that he is cautious around new things, don't walk into the room wearing a motorcycle helmet or Halloween mask. You might start with the new item on the ground. Your puppy's body language will be the best barometer on how to proceed.

Puppies and Children

We had a dog when I (Marge) was a kid. It was a magical bond, from my point of view, from the very beginning. My first dog, Nikki, was a white and brown hound mix and she's the reason I've loved dogs as long as I can remember. I vividly remember her standing on top of a snow fort we built, her tail happily wagging.

Eileen here. When I was growing up, my family had dogs, too. We got a good-natured beagle puppy when I was 3 years old, and I spent a lot of time with him. It was a good thing he was good-natured; I was too young for the things we did without supervision. My dad also had a hunting dog. But "my" first dog was a little schnauzer-ish mutt who appeared in our irrigation ditch when I was 12. I begged to keep her, and we did.

Marge and I were lucky with the dogs who came into our families. But we don't want *you* to leave it to chance and luck. As we've mentioned before, back then, a lot of dogs got socialized by default because they had more freedom in their lives—especially dogs in the suburbs (Grier, 2010). The flip side was that they were less supervised and in more danger. The sweet beagle I grew up with? He got hit by a car and died when he was about 11. I'm sorry to tell that sad story, but it was such a common fate when dogs had more freedom. I support the efforts we make now to supervise and contain our dogs for their safety. But we also have to be much more proactive about socialization. We don't want you to depend on luck.

Even though this book is primarily about socialization, we think that living successfully with puppies and children is important enough to widen our focus for a bit. So let's talk about puppies and children and how you can set up for successful interactions generally. Along the way, we'll point out how this fits into your puppy's socialization process. Then we'll talk about people who don't have kids and the steps they'll need to take.

Bringing a Puppy into a Home with Children

Young puppies who go home to live in families with children often do pretty well, as long as the children are supervised and taught appropriate behavior with dogs and puppies. From the day he comes home, the puppy is exposed to kid noises (squeals, laughter, crying), toys (flashing lights and noises), and running and jumping. Plus, kids often carry (and drop!) snacks. Living with children usually helps puppies form positive associations with children.

That said, parents almost always underestimate how much work a puppy is. The to-do list includes socialization outings, near-constant supervision, handling, training, classes, exercise, and vet visits. All those things should happen in the puppy's SPS and well through his first year. If you don't have time to properly expose your puppy to life in a human world and train him, please consider delaying getting your puppy until you do have time. If you already have your puppy, create some time in your schedule so you can give your puppy what he needs now. This will help him, and you, later on.

The best approach for parents will depend on the ages of your children. Toddlers and puppies actually have a lot in common. They both need tons of supervision! We recommend making extensive use of what Jennifer Shryock of Family Paws Parent Education refers to as Success Stations. Success Stations are places where your puppy can be successful, because they provide the puppy plenty to do but prevent direct physical interaction with the child. We link to a blog post about them from Family Paws at PuppySocialization.com/resources. Puppy exercise pens, crates, baby-gated areas, and indoor tethers (use only if your children are too young to be mobile) are examples of Success Stations.

Plan ahead and have the basics set up before you bring your puppy home. Then you can make adjustments as you get to know your puppy and learn what works best for him and your kids. I (Marge) don't expect a toddler, child, or teenager to be able to comply with my instructions 100% of the time (close the gate, leave the puppy in his crate, don't bother him while he's eating/sleeping). Heck, I can't get my husband to be 100% compliant with the things I ask him to do with the dogs, and he's an adult! Expecting 100% compliance from children is not realistic. And we also don't expect your tiny baby puppy to be able to control himself around your infant or toddler. That's not realistic either. He's an animal. With really sharp teeth. And he's got instincts that tell him to chase things that run, bite to play, and protect himself if he feels threatened.

Babies and toddlers don't generally have the physical or developmental skills to follow your admonitions to "be nice." And your number one priority with puppy, baby, and toddler interactions is safety. Have a plan. Manage and monitor. Have your puppy in a Success Station, like a puppy playpen, provide him with a long-lasting food item or toy, like a stuffed Kong, and have your baby or toddler nearby. The puppy is safely confined, he has something yummy to help him form positive associations, and puppy and baby are performing an age-appropriate activity: parallel play. They don't have to be directly interacting with each other to form positive associations with each other. We could write a book on dog and baby safety. We don't have to. There are some wonderful books and resources already written. They are included on our list at PuppySocialization.com/resources. And if you choose to consult a professional, a licensed Family Paws Parent Educator can help set everyone up for success.

Children Helping with Training

We have said more than once that this is a book on socialization and not on training. At the same time, some basic tips on how kids can interact with dogs through training can go a long way toward a puppy's success in the world. I (still Marge) love it when the whole family is involved with training. Everyone is training the puppy each time they interact with him whether they realize it or not, anyway.

Involving children in positive reinforcement training teaches them empathy, how to behave so their puppy behaves, helps them bond with and develop a positive relationship with their puppy, and teaches them how animals learn. Those are some pretty good life lessons from something with big eyes and soft fur.

All of that said, it's important to know that behaviors that are normal and natural for kids make dogs uncomfortable. Some children are close to eye level, toddlers walk like T. rexes, high-pitched kid voices get dogs excited, and kids run, jump around and play—all very exciting to dogs. So the more the whole family can learn about dogs, their behavior, and their body language, the better prepared they are to separate them from the storybook myths that we all learned as kids. This keeps children safe and keeps dogs in homes. As you learn these things, don't keep the knowledge to yourself. We hope you'll share it with every child you know.

Older children can be training helpers depending on their age, maturity, and level of interest. Children can learn good training skills very quickly. But keep in mind that **even mature children are still children**. The adults are responsible for the training and development of the puppy. It is up to parents to teach children how to be respectful of dogs and puppies. It's also up to the parents to teach children to recognize when dogs or puppies need some space.

During training, we recommend young children put treats on the floor instead of delivering them to the puppy's mouth. Puppies can get excited and possibly jump at the food, and that can scare or hurt young children. The children might then hesitate or deliver the treat so slowly and cautiously that the pup jumps and lunges for the treat more! It's not a good cycle. If children always put the treats on the ground, that pulls the puppy's focus downward. Food on the floor also helps him keep four feet on the floor. You can see the "assistant trainer" in the next video demonstrate this technique. She's using a clicker, too!

Video 5.6 A child clicks and reinforce the family's puppy, putting the food on the floor

I (still Marge) teach children how to "Be a Tree"—stand still, fold their branches (hands together in front of their bodies), stare at their roots (feet), and count in their head until a parent can help. They should ignore the puppy, and be still like a tree when their puppy gets too excited. This is a technique from Doggone Safe, an organization that provides educational materials for families on dog safety. They are worth checking out (linked at PuppySocialization.com/resources). I also teach puppies to sit automatically when kids stop moving.

The children in the following video are developing a strong bond with their new puppy. They are also teaching the puppy how to behave in a way they like instead of jumping and biting. Don't the kids do a great job?

Video 5.7 Children reinforce their puppy for polite behaviors

Growing up in a home with a dog is one of the special gifts a parent can give to a child. And puppies who are exposed positively to kids have an advantage in life, too. (Arai et al., 2011). Dr. Sophia's Yin's handout, "How Kids Should and Should Not Interact with Dogs," linked at PuppySocialization.com/resources, is a great resource. There are so many wonderful parent-guided activities children can enjoy with dogs. Dr. Yin recommends the first four of the following activities. We added a few of our own.

Fetch. Fetch is a great game for kids and dogs to play. But caution must be used with puppies with any repetitive activity. This is because the plates of cartilage (growth plates) at the end of the pup's long bones have not yet grown, hardened, and fused to the bone (Von Pfeil & DeCamp, 2009). Ever wonder why puppies are so floppy and squishy? One reason is that they have spaces between their bones and joints. The plates are easily damaged, so you shouldn't ask puppies to perform repetitive activities for long at all. You can find charts to tell you at what age your breed of puppy's growth plates usually close. In general, the smaller the breed, the sooner the plates close.

I (Marge) like to teach puppies to drop the ball in a basket or container. That way, the children don't have to take the ball directly from the puppy's mouth. Children should not take toys or balls (or food and bones for that matter!) away from dogs and puppies. Remember when we talked about teaching children to be respectful of dogs' space? That applies to dogs' things, too (toys, food). One way to build good habits for both the kids and the puppies is to teach the pups to drop things. The adults should teach the puppy that skill before the kids play fetch with the puppy. Teaching them is the adults' job and is another socialization opportunity, because a basket is a novel object! In the following video, you can see a lovely example of this kind of fetch game. The adults started the training and as the puppy gained the skills, the oldest child took over, with supervision. My "assistant trainer" in the video is 11 years old.

Video 5.8 Apple plays a child-safe game of fetch

Tricks. Teaching tricks is a great activity for kids and dogs. Some of our favorites are hand targeting (puppy's nose to person's hand), spin (both directions), riding a skateboard, and paw targeting (puppy taps an object like a noise-making button or a margarine tub lid with his paw). Be sure to first introduce any objects you need for tricks by using what you learned above about socializing your puppy to novelty.

Walking and running. Walking and running are great activities with the puppy on leash and when supervised by an adult. Young children should not walk dogs alone. It is not fair to others walking dogs if your child can't control your dog. If your dog or puppy gets excited at the sight of a person or another dog and runs up, that is not safe for your child, your puppy, or the other owner. Just as with playing fetch, walking and especially running with the puppy need to be limited in time and intensity so their growing bones can be kept safe.

Hide and seek. This is one of our favorite games to teach puppies and kids. It is also a good building block for "come when called." The puppy gets rewarded when he finds the "hider."

Find the treats. Finding treats is a fun, outdoor game (you don't want to teach the dog to search for food inside). Get some high-value, stinky treats. Keep your dog indoors or in a sit-stay if you've taught that, and help your child place the treats in a small area outside. Then release the puppy to find the treats. Help him the first couple of times if he needs it by walking toward the area with the treats. He'll learn fast! As he learns the game, you can make the search harder by spreading the treats out or moving the treats closer to objects like a flowerpot or leg of a chair. Pretty soon you'll be able to hide the treats *behind* objects, such as the leg of the chair or on the other side of the flowerpot. Always make sure to use safe hiding spots. If your child doesn't know what that means, make sure to supervise.

Find your toy. This is a fun game for puppies and kids. Play with your puppy and get him interested in a toy. Keep your puppy with you and have your child place the toy a few feet away from the puppy and return to you. Release the puppy to the toy. When he gets there, you can either play with him with the toy or reward him with food. Repeat, with your child gradually moving the toy a little farther away, just out of sight, and as with the "find the treats" game, gradually work up to hiding the toy. The goal is to build a lot of confidence with this game. The adults trade the puppy a cookie for the toy and the child gets to "hide" it again with direction from the adult.

Potential Flashpoints Between Puppies and Children

There are some things you need to be careful about when managing kids and puppies. If you paid close attention to our suggestions above, you can probably guess the kinds of problems we were steering you away from. Here they are.

Puppies have really sharp teeth. It can be hard enough for adult humans to cope with sharp puppy teeth. It can be extra hard for children and even dangerous for toddlers or babies. Puppies and young children must be actively supervised and be physically separated when you can't supervise them. As we said, you can't expect a child to always remember to interact appropriately with the puppy. It's also unrealistic to expect the little puppy to not do puppy things, like jump and bite with sharp teeth. Again, puppies and kids need supervision.

I (Marge) wanted to show you what Zip did to one of my shirts. He actually did this to a few of my shirts. This is a good reminder that puppies can be too fast even for experienced adults. Your puppy could easily injure a child. Supervision is a must.

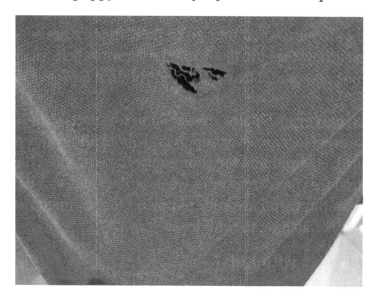

Kids get puppies excited. Kids, especially kids under the age of 9 or 10, do things that puppies find exciting. Kids run, squeal, climb on furniture, and move in unusual ways. All those things can prompt the puppy to chase and often he will jump and bite at clothes.

Kids and puppies show affection differently. Kids (and adults) show affection by hugging, kissing, and making eye contact. Guess what? Puppies and dogs often hate those things. Eye contact is natural for humans. It's a way we show love, connection, and interest. From the time a baby is born, we make eye contact. Hugs and kisses are also connected with affection for humans. But none of these things is popular with dogs. When a dog makes direct eye contact with another dog, it's often a threat. Puppies and dogs might consider direct eye contact from a person as threatening or at least unsettling.

You need to teach kids that dogs don't like to be looked straight in the eye, hugged, or kissed. In about 80% of dog bites, the dog belongs to the victim's family or someone they know (like a relative, caregiver, friend, or neighbor). These aren't wild packs of dogs running around biting people. They are your dog and your neighbor's dog. And the group on the receiving end of those bites most often is boys ages 5–9 years old (Centers for Disease Control, 2001). Even if you have the most tolerant dog in the world, it's important to be a role model and teach your children how to interact with dogs appropriately. Because when they go to their friend's house, how they interact with their own dog is how they are likely to interact with their friend's dog. And their friend's dog may be nervous and fearful around children he doesn't know.

Prepare for caregivers. Finally, please don't expect your child's caregivers (babysitters or relatives) to manage or supervise your children and your puppy. Set up a safe, but contained, environment for your puppy to be in when the caregiver is with your children. Introduce your caregiver as we described above as part of your puppy's socialization process. But then have your puppy or dog confined with a yummy chew when your caregiver is watching the children.

What If You Don't Live with Kids?

Marge here. Since my husband and I don't have kids, I had to go out of my way to expose Zip to children when he was a puppy. Still, he is definitely not as used to young human movements, noises, mannerisms, or crying as he would be if he lived with a child. Nor is he used to humans having their own stuffed toys, as you'll read about below! So when we are around children, I am extra careful to manage the interactions. Several interactions with "kids" as a puppy did not prepare him for the myriad of interactions he could have with the age group of humans 0–13 years old.

Children move very differently from adults, they have higher pitched voices, and they are often "magnetically" attracted to puppies. One study shows what happened to puppies who did not form positive associations in their SPS with children. When the puppies encountered children later in life, they showed increased heart rates and were more likely to aggress toward the children than the puppies who did have positive exposures during their SPS (Arai et al., 2011).

If you don't have any children or your children are now adults, you'll need to go out of your way to expose your puppy to children. You might think, "Why bother?" The reasons to bother are friends, neighbors, family, the veterinarian's office, agility class, walking down the street—your dog is likely to encounter children at some point during his life. Do him a favor and include these little humans in your socialization efforts while your puppy is in his SPS. Control and manage the interactions, just as you would with an adult; there are just some extra considerations when kids are involved.

My nephew (still Marge!) and his family came to visit us. It was a wonderful visit and we all had a great time. Unfortunately, Zip and Liliana (my grandniece) got off to a rough start when Zip stole her baby doll. In fairness to Zip, up until that point, any stuffed toy in the house belonged to the dogs. We set up a structured interaction between Liliana and the dogs and used the stairs to create a barrier and some space to help them feel safe.

Video 5.9 Marge's dogs and young Liliana build positive associations with each other

Given the COVID-19 pandemic, sometimes social distancing will require some creativity. Eileen here. When Clara was a puppy, we lived a block from a residential day care. We would go for a short walk and *listen* to the kids who were in the fenced back yard of the day care. There was lots of shrieking for Clara to hear and get used to. Do you think I used food to create positive associations? Bingo! Opportunities to observe kids from a safe distance are great, too, but be careful you don't give people cause for alarm. I was lucky; I knew the daycare owner so she knew why I hung around on the sidewalk sometimes.

Planned Sound Exposures

Your puppy will be learning about the sounds of a human household from the moment he comes home, or even earlier if he comes from a good breeder or foster home. If he's in his SPS, he will probably take many sounds in stride. He will habituate to them and they will blend into the background of his life.

You can also stack the deck in your favor with some planned sound exposures. You can help your puppy get used to different sounds and banging noises by starting with low intensity sounds and building from there. You may have heard of deliberately dropping pots and pans in the kitchen as a "socialization" technique for puppies, but that is way too startling and intense a starting point. But dropping something smaller, lighter, and non-metallic onto different surfaces and from different heights will allow you to start small and work your way up. Here is a simple plan that starts with muffled noises that gradually get louder.

Get a plastic or firm paper cup (not Styrofoam). Set yourself up to be able to drop the cup from different heights and onto different surfaces. Here is one way to proceed.

With your puppy nearby (but not right next to you), drop the plastic cup onto:
1. Carpet or grass from about a foot high
2. Carpet or grass from two feet high
3. Hard flooring from about a foot high
4. Hard flooring from two feet high

Continue this pattern, then switch to something noisier, like a small metal lid, and start back at the beginning.

Each of these steps would be followed by a treat of course, or a game the puppy loves. And proceed very slowly and watch your puppy. The point of the exercise is for him to habituate to sudden noises. If you accidentally scare your puppy, stop for the day. Start another day with a quieter object and a softer surface.

This process is good for any puppy, but extremely helpful if you plan to play agility. Agility teeters make a banging noise that can be quite loud, depending on individual equipment and the weight of the dog. You are giving your puppy a gift by making noises predict great things. You can use this principle for all sorts of noises. Some brilliant agility trainers would make that even better by having the puppy knock the objects to make the noise (Vegh & Bertilsson, 2012, p. 175–179). The point is, even with sounds, you can control the intensity.

The next video shows Marge introducing Zip to a grooming dryer in her home. This is not Zip's first exposure to a dryer. He had one used on him when he was still with his breeder. Clipper and dryer sounds are extra important for breeds that will go to the groomer. Working on these sounds at home can help put some deposits in the "groomer" bank account before you ever get there!

Video 5.10 Zip experiences the grooming dryer at home

Eileen here. There is a lot to learn from this video. First, Marge introduces Zip to the dryer without power or sound. Second, when she does turn it on, she leaves it on only for only a short period, about 10 seconds the first time. If your puppy has never been exposed to a dryer, an even shorter exposure, such as 2–3 seconds, would be a good starting point.

Finally, here is a really important observation point. Fearful behavior can take many forms. After the second time the dryer makes a sound, Zip starts to vocalize and bounce around, making movements that resemble play bows. But notice, as Marge did in the moment, that he's staying behind her. He is over-aroused and stays upset for almost a minute (that's forever in puppy time!). She is able to use a simple trained behavior to calm him.

But the most important thing is what Marge doesn't do. Spoiler alert: she doesn't turn the dryer back on after Zip calms down. She saves that for another day. And when she does, she makes sure he is comfortable with the dryer on its own, then starts with a shorter sound. Zip did great. Dryers and their weird sounds became an unremarkable part of his world.

We've talked about getting your puppy used to typical noises in a graduated way. Not every puppy is typical. If your puppy reacts to certain sounds with trembling, drooling, or other fearful body language, and particularly if it takes him a long time to recover, consult your veterinarian or a veterinary behaviorist now. Sound phobias are rare in puppies, usually cropping up when the dog is a young adult or older. But at any age, according to Dr. Karen Overall, noise phobias (as well as storm phobias) in dogs are medical emergencies (Overall, 2004). We link to her excellent article at

PuppySocialization.com/resources. We can't ever assume noise phobias will go away on their own.

Unexpected Sounds and Events

Throughout this chapter, we've described how you should plan for and set up socialization for your puppy at home. Now, let's plan for the unexpected.

Sudden sounds can be one of the most startling things for animals to experience. All mammals with hearing startle at sudden, loud noises. Your puppy will, too. So what should you do? By now, you know the drill: pair with food and play.

If you accidentally drop a heavy pot in the kitchen, what do you do? "Yay! A sudden noise! That means treats!" Then give your puppy several treats in a row. Remember classical conditioning? How a neutral or scary event can begin to predict something great? Teach your pup that sudden sounds predict treats raining from the sky or his favorite game. As Marge suggested earlier in the book, if you keep food in one pocket and a toy in the other at all times, you'll always be ready.

You should do the same thing for all startling events, whether they include loud sounds or not. Think of this scenario. You are sitting at your desk and your puppy is lying on his mat next to you, chewing a chew toy. You reach up to the bookshelf and pull down a book, but you accidentally knock two other books off the other end of the shelf. They fall on the floor and startle your puppy.

Party time! "Yay! Books fell! I get to give you treats!" You whip out your treats and start feeding that puppy. Feed, feed, feed. And also play with him. "Isn't it great, pup? Some books fell down over there, and that means you are getting awesome treats! Wow! Now would you like to play a game for a minute? Shall we tug on a toy together? Throw a ball?"

This routine is intended for minor accidents and other startling events. Things like a sudden knock at the door, a delivery truck in the driveway, a dropped baking sheet; you get the idea. If your puppy is petrified and won't eat, don't force anything. Just let him move out of the area, and try to give him treats further away from the event. Check our section on dealing with abnormal fear in the SPS. If your puppy startles at a lot of noises, we want you to see your veterinarian.

Things humans take for granted might be startling to a puppy. The puppy sees mail tumbling through the mail slot for the first time. Yay! Treats! Or maybe the curtains suddenly move when the air conditioner comes on. Here come the treats! With enough practice, your pup's response will likely generalize so that he isn't afraid of other startling events in life. The goal is a puppy who, instead of freaking out or creeping away when something startling happens, will calmly look up and say, "Hey, do I get a treat for that?" (Hint: give him one.)

Notice the difference between the two processes we have described: planned exposures and surprising events. When working with a neutral object, like the soup can we suggested earlier, you wait for the pup to interact with it in any way, and you give treats for that. With unexpected (for the puppy) changes in the environment, you are working on the fly. You quickly follow the incident with treats "raining from the sky." Your pup doesn't have to do anything. You simply pair the startling event with yummy food. This will become habitual for you after a while. Even when you have been startled too, your habits will kick in so you can help your puppy.

Handling

Your puppy will need a lifetime of handling—by you, the veterinarian, perhaps the groomer, friends, and children. Everyone will want to touch your dog. Some handling is necessary to maintain your dog's health and some of it will just happen, because that's what people do.

Humans show affection through touch, and sometimes (unfortunately for dogs) through eye contact. If your puppy was born in the care of an experienced breeder or foster family, he started getting handled as soon as he was born. But even if that happened, your work isn't done. You've still got plenty to do. And if your puppy was not handled regularly, you have some ground to make up. Your time investment now, in the sensitive period for socialization, is much smaller than it will be if you wait to tackle this when your puppy is older. And if you're like most people, you're going to handle and interact with your puppy anyway. Why not make sure to check all the boxes (plan!) and help him form positive associations with the types of handling he'll experience throughout his lifetime?

You'd be surprised how many dogs are not prepared for handling at the veterinarian. Marge here. Many of my clients are referrals from veterinarians. Veterinarians refer them to me for behavior modification because routine veterinary care has become stressful for the dog, and sometimes unsafe for the veterinary team. I've had client dogs with the following problems:

- One had to be sedated for routine ear care (and still struggled under sedation);
- One got stress colitis after grooming;
- Many dogs struggled and became extremely distressed for nail trims;

- One could not have blood drawn and as a result was not on heartworm prevention; and
- One aggressively cornered the vet while barking and lunging, and could not be groomed or boarded.

You get the idea. For all these dogs, their owners thought they were "fine" because they weren't growling, biting or actively resisting. And then "all of a sudden" they couldn't be handled any more. And we want to remind you once again of the importance of reading your dog's communication through body language. These dogs probably did give signs of discomfort before they escalated into panicky behavior or aggression, but their owners, and possibly the professionals handling them, missed the signals. So let's take action now to make sure routine handling by professionals, family, or friends is not stressful for your pet.

How can you do this? You can probably guess what we're going to say. We want you to make handling a predictor of food. Yep. You can use your puppy's regular meals for this. You can hand-feed during handling.

Remember, for classical conditioning to occur, touch/handling must come before the food. That said, puppies are wiggly and use their mouths to grab everything. With puppies, I (still Marge) often start with management (food first and then handle) and then switch to handling first, then food. There will be times when you must do something, like trim your puppy's nails, before you've created a strong positive emotional response to that particular type of handling. Managing the puppy with food is a good way to get your task accomplished and keep your training on track.

Using food to manage your puppy avoids wrestling matches. These don't turn out well for anyone. Before you think, "But, but, I had to apply eye drops, administer ear medication, trim his nails, brush him," or whatever, I'm here to tell you there's a better way. I see the aftermath when puppies are wrestled to the ground for eye drops, ear medication, nail trims, or brushing. Those puppies often become avoidant and sometimes aggressive. There are ways to get those things done without a wrestling match.

In some cases you might need a professional trainer to coach you through the process. For instance, you may be able to handle nail trims, but for eye and ear drops and more intrusive things like wound treatment, it's probably best to get help. But we can show you the basics, and often you'll be able to make great progress yourself because it's the same principle we've been following. We can use food to make handling easier for the puppy and ourselves.

The following video featuring my dog, Zip, when he was a puppy shows the process I use for introducing handling for eye drops. Some people might take exception to Step 1 in the video because it does not change how the puppy feels (remember how important the order is in pairing?). But I find it helps prepare the puppy for the areas that will be handled.

Video 5.11 Zip gets introduced to handling for eye drops

The trick with handling is to start at a level the dog will accept without withdrawing from you. For example, if your puppy doesn't like to have his feet handled, don't start by handling his feet. Let's use the goal of handling feet as an example, though. Here's how to proceed.

Start by touching the puppy in an area he likes and then give your puppy some food. That way, touch in any location predicts food. Do several "easy" touches (touch he likes), followed by food, then touch or scratch his shoulder, and follow that with food, too. Then go back to another place you know he likes (like under the chin, on the chest, behind the ear). Gradually, in tiny increments, work your way down his shoulder, incorporate sliding your hand under his leg. Pause with your hand under his foot, add your thumb to the top of his leg, pausing at his paw, and gently squeeze his toes to extend the nail. Each touch would be followed by food.

Those types of more "difficult" touches (handling the puppy is uncomfortable with) would be sandwiched between types of handling the puppy likes. I always progress at the puppy's pace and will make sure to take fun breaks before the puppy gets fussy.

Getting your puppy used to nail trims will involve handling their feet as well as getting them used to the sight and sound of the nail clippers.

Some puppies may not like the sound of the clippers when they clip. You can separate out the sound by clipping a piece of dry spaghetti near your puppy and following that with food. (Don't feed him the raw spaghetti, though!) Clipping spaghetti conditions the sound separately so when you *do* clip a nail, the pup is experiencing only one new thing, the sensation of the clip. He will already think the sound is great.

The following video shows the progression for one puppy. Your puppy might progress at a faster or slower rate. Every puppy is an individual and it's important to adjust your training plan for the puppy in front of you on that day. Notice, though, how we follow the first successful clip with several easier touches. It's human to want to charge ahead after a successful clip and immediately clip the next nail. "Hey, he handled that great—I'm going to go ahead and get another one done!" We have all done it, but that's exactly what not to do. If you push, you are likely to lose ground.

Video 5.12 Puppy Taylor progresses from fear to acceptance and enjoyment of nail trims

In the following video, you can see puppy Zip's reaction to being brushed. His response of wriggling around and trying to bite the brush was very typical—and most puppies I know would do the same thing. But in the second clip in the video, the brushing goes much better. Sometimes you can have brushing predict food. That is good training. Brushing predicting food helps him feel better about brushing. Other times you'll just need to keep his mouth busy so you can get him brushed. We call that management and it is fine to do that. Just make sure you do plenty of the former too.

Video 5.13 Marge uses food to give Zip something to do while being brushed

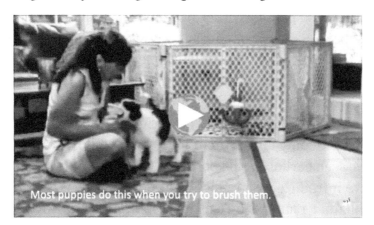

Most puppies do this when you try to brush them.

There are lots of other handling categories and Dr. Yin's socialization checklist (linked at PuppySocialization.com/resources) covers many of them. Remember, the goal is to avoid startling, scaring, or hurting your puppy. The goal is to get him used to the types of handling he may experience during his lifetime. It is better to start with a small approximation of the behavior (touching his body) than to suddenly subject him to a behavior (grabbing his foot) that hurts or scares him.

Other Dogs at Home

The approach when introducing your puppy to a dog you already have will depend on the particular dogs, but a good rule is to take it a lot slower than you think you have to. Think about what you will need to set up if you are bringing home a puppy of a large, gregarious breed and your resident dog is a senior toy breed. Or maybe your current dog is one of those fantastic "uncle" dogs, and loves to raise puppies. These dogs give puppies what trainers refer to as a "puppy license," because they let puppies do all types of things to them that they would never tolerate from an adolescent or adult dog. They gently teach the puppy a variety of subtle communication skills to learn when interactions are welcome and when they are not. Your approach will be different in those situations.

In every case, you need to make sure the resident dog doesn't scare or harm your puppy (it can happen). In turn, you need to protect your resident dog from annoying puppy behavior.

Barriers and management are your friends. Not only do barriers make life easier for the humans, but they also help the other dogs living in the household. I (Eileen) kept my puppy Clara separated from my short-tempered, middle-aged dog Summer using the French doors that split my house in half. I separated them for the first couple of months I had Clara because I wasn't sure Summer would grant Clara a puppy license. I also separated Clara from my elderly, frail dog Cricket because Clara was growing fast and was clumsy and rambunctious. That's right: I had four dogs in the house and I let Clara hang out with only one of them for quite a while. I was lucky: my other dog Zani, whom you've seen earlier in this book, was a born puppy raiser. But if she hadn't been, I would have introduced Clara slowly to her as well. While my situation might not be typical—I hadn't planned on getting a puppy; Clara was a feral puppy who wandered into my house—it still pays to take things slowly.

Take the time to do it right. And when you think you are going slowly, go slower still. Remember, you want to be sure your puppy forms positive associations during their SPS, and that includes learning about any dogs in your home.

We have a great visual example for you. In the next video, Marge shows how she gradually introduced Tinker, a fox terrier puppy she was boarding, to her dog Zip. I (Eileen) will walk you through some highlights.

Zip has always been friendly and gregarious and was likely to enjoy the pup, so Marge didn't have to worry about aggression. But Zip was an adolescent himself and needed some time to calm down and learn to be gentle with the tiny puppy. Look at the size difference! He was an exuberant greeter, and that's putting it nicely.

Marge gave little Tinker her own space in an exercise pen, with food, water, a bed, and toys. She kept Tinker inside the exercise pen and carefully controlled Zip's behavior toward the puppy for the first couple of days. Tinker was already a little worried about being in a new place. Zip was very curious about the puppy, but Marge did not allow him to charge up to the pen off leash and scare or startle her. She kept Zip on leash when he was near the pen. Remember, to a tiny little puppy like Tinker, Zip must have looked like Godzilla.

In the video, you can see the setup and how the dogs settled over time.

Video 5.14 Teenage Zip and tiny Tinker have positive, controlled interactions with each other

Marge chose this setup for these two particular dogs. For dogs you are less sure about, you could start with them completely out of sight of each other, then with an "airlock" area between them.

Putting the two dogs adjacent with a fence between them was a good method for Marge's friendly, near-adult dog and the confident puppy. But there are many situations in which it would not be appropriate. Here are three of them:

- If you have a grumpy, snarly mature dog (the last thing in the world you want to do is park him next to a puppy with only a wire fence between them)
- If you have a large breed, exuberant puppy who would enjoy bouncing against that fence and a tiny, fearful, or frail adult dog
- Any dogs you don't know, unsupervised, no matter how well they are apparently matched

So put some thought into your setup. Your puppy and your resident dog or dogs will need separate spaces at least part of the time. If you have a senior dog, don't let your puppy pester him. Intervene and give your senior a separate space if he needs it.

Keep in mind that your resident dog or dogs and your new puppy will probably be spending years together. They don't have a choice about it; you are bringing them together. Take the introductions slow and easy. Give them time to view each other from a distance, and to learn each other's noises and odors. You can even trade their bedding around a little. If you think your small breed puppy will get trampled by your young Labrador, keep them apart for a while. Then keep the Lab on leash, as Marge did with Zip, when first allowing the dogs in the same space.

Since Tinker was a client's puppy and did not stay long, Zip never did get to play with her off leash. He was too clumsy and goofy (did you see the paw to her head?). He did learn a lot though, including a softer approach and play style. But you know the drill: the paramount concern with a puppy this age is providing positive experiences.

Being gentle with a puppy is not something a human can directly teach a dog, but Marge facilitated it with controlled exposures and lots of breaks in the play. I know she is counting her blessings that between her efforts and Zip's natural friendliness, he learned gentleness through direct experience with the puppies themselves.

Now, would you like to see what it looks like to introduce a youngster to an "uncle dog," one of those great puppy raisers?

The adjacent photo is from the day Marge's Rhodesian ridgeback, Chase, and the puppy Rounder meet for the first time. Shortly after they met, they were playing together. Chase was about 90 pounds when he met Rounder. He made himself less threatening by lying on the ground and invited Rounder to play. Rounder accepted the invitation and that was the first of many years of wrestling matches. What do you see that indicates the pup and the dog are playing and not fighting?

In the next photo, the student has become the teacher—Rounder, as an adult, invites puppy Pride to play. Notice the soft, loose muscles, and the soft eye. This invitation to play started a game of chase, and these two continued to play together for years. It was lovely to see Rounder pass along the lessons he learned to the next generation of family dogs.

Adult dogs who are good puppy raisers are worth their weight in gold. They teach puppies the subtle art of canine communication—and also to use their mouths softly. If you're lucky and have a well-trained resident dog, they will also help by modeling good behavior for your puppy, such as housetraining and coming when called. Rounder learned to ring the bell to go outside by watching Chase. Marge hadn't started that lesson yet with him and he started offering the behavior just by observing it.

Here's one more example. I (Marge) do a lot of management when a puppy plays with another dog. And by management, I mean I actively watch them as intently as I'd watch a toddler by a pool. I also interrupt the play often. The puppy in the next video belongs to one of my clients. Watch the video to see some of the things I do to facilitate a good play relationship between the puppy and Zip.

Video 5.15 Zip plays appropriately with a younger puppy

If you are fortunate enough to have a resident dog who is good with puppies, count your lucky stars. But don't stop there. As far as the COVID-19 pandemic permits, your puppy should meet a variety of dogs of different appearances, breeds, and ages throughout his first year of life.

Spending Time Alone

This may seem like an odd thing to put in a book about socialization, which is all about exposure to new things, right? But being alone probably *will* be a new thing. It's possible your puppy may have never been alone before. First, he was with his siblings and mom after he was born. Then he left everything he knew and came to live with you and accept your family as his new family. Humans come in and out of the home, and dogs have to stay home alone sometimes. And if you got your puppy from a situation where you know less of his history, he might have experienced traumatic time alone even before he came to you.

How many people with good intentions and love in their hearts for their new baby pup take a couple of weeks off work when they first get him? There's nothing wrong with that in itself, but what happens when they suddenly start disappearing for hours at a time after their vacation?

By now, you know what we are going to say. You need to expose your pup to time alone gradually, just like anything else that is novel for your puppy. Take those two weeks (or more) and use them to introduce alone time for pup. And you need to work up to absences as long as your planned absences will be, also keeping the puppy's age and capabilities in mind. (Puppies shouldn't be left alone for long. We'll ask you to check with a good puppy training book for that information.) Imagine the stress on both of you if you head to the office and suddenly leave the puppy alone for longer than he's ever been before!

But arranging some time alone is important even if you work from home or are sheltering in place because of the COVID-19 pandemic. As a trainer and behavior consultant, I (Marge) regularly see the problems faced by "pandemic puppies" when their people are no longer home 24/7. The effects can be heartbreaking. These puppies may have spent months, rather than a week or two, with their owners always present. Even if someone is *always* home in your household, spending time alone is still a necessary skill for your puppy. It may be even more important! Situations change. In the 10–15 years of your dog's life, he'll probably need to be alone now and then.

Safe Confinement Area

Create a comfortable and safe confinement area for your puppy using an exercise pen and baby gates. Exercise pens and baby gates are worth their weight in gold and I hope you invest in some! I like to think of this as creating a playpen for your puppy: a safe place for him to be when you can't give him your full attention. Your little baby puppy doesn't know he could die if he chewed on an electrical cord (that happened when a friend was babysitting a litter of puppies). He also doesn't know the difference between the toys you buy for him and your own stuff. Part of your job is to keep the puppy safe as he learns about life with you.

As a professional dog trainer with thousands of hours teaching dogs and their owners, I still wouldn't be able to properly keep track of a puppy who had access to my entire house. I don't ever recommend this to clients. The confinement areas make it easier for the human to keep both the puppy and their own belongings safe (not to mention helping with housetraining!).

This safe place is also where you can start with alone time training. Make the area comfortable and attractive for the puppy with a soft bed and toys. If you plan to crate train your puppy, you can put the crate in the pen and fix the door open. Drop some tasty treats in the crate and maybe even scatter some in the pen. Feed your puppy in his confinement area to create more positive associations. We have a video linked at PuppySocialization.com/resources showing puppies learning to love their crates by being fed there. Crate training is beyond the scope of this book, but if you choose to do it, the principles of "start small and build" and "create great associations" you are learning here will be integral to that training.

The following photo shows Zip choosing to settle in his crate on his own in his confinement area.

Practicing Alone Time

1. Before ever confining your puppy, put the most comfortable beds and most fun puppy toys in the confinement area. Feed your pup in there, too. Remember classical conditioning? You are using it again!

2. Make sure his physical needs are satisfied (food, water, exercise, and elimination) before putting him in his confinement area.

3. Start with short duration separations.

4. Make it easy on him and yourself by first putting him in his confinement area when he is already tired and ready for a nap, or ready to settle down with a chew toy he loves. Then put him in his pen and go about your business. He may not even notice the confinement or your brief absences.

5. The first few times, you can hang out in the same room as your puppy until he falls asleep. Just like babies, puppies sometimes fight sleep when they are tired. Most puppies generally settle down after several minutes.

6. Your puppy may vocalize (bark or whimper) at first. His time in his pen alone may be the first time he's been physically alone in a space, so being mildly uncomfortable is normal. That's why we've set him up for success by making sure all his needs, including exercise, are satisfied before we put him in his pen.

7. Try not to stare at the puppy when he is in his pen. Looking at him starts an interaction with him. If you can't resist watching, there are baby monitors and baby monitor apps for your phone.

8. Gradually increase the time of the separations.

9. Try to get your puppy out of his pen before he wakes up (generally about an hour) and immediately give him an opportunity to eliminate. If he has been chewing rather than sleeping, get him out about the time he finishes but before he starts to fidget.

10. Make it part of his routine. If you work out of the house, practice separations even on days when you are home.

11. If your puppy doesn't settle down after several minutes, or if he is panicked or distressed, take him out of his pen. You may have misjudged that his physical needs were satisfied. **Do not make a distressed puppy remain in his pen. Do not try to make him "cry it out."**

12. Try again at another time. If he shows the same signs of distress, consult your veterinarian or a veterinary behaviorist.

Using this kind of setup can organically incorporate micro-separations throughout the course of the day—exactly what your puppy needs for practice. He doesn't need to follow you into the restroom, and he needs a safe place to be when you take a shower. He will also be learning that you leave and come back several times throughout the day—a very good thing for him to experience.

Think of alone time as independence training. We want our dogs and puppies to feel safe, even when we are not there. A "Velcro dog," one who is stuck to your side 24 hours a day, 7 days a week, is a dog who hasn't been given the gift of feeling safe when he is alone.

Again, if your puppy shows distress at being alone and doesn't improve, this can indicate a serious problem. You should consult with your veterinarian or a veterinary behaviorist.

Chapter 6. Socialization Away from Home

Here it is: the part people traditionally think of as "socialization." You are going to take the puppy out, show him stuff, and introduce him to people. But you'll be prudent about it. You are learning how to read your puppy's body language. You are learning it's your job to be his advocate. You are learning to curate good experiences for him. And you know he needs more than mere exposure, so you are planning ahead and will be ready with your treats, toys, and anything else you need to bring (like a mat or dog bed).

When to Start Outside the Home

Are you supposed to wait until you finish your list of "home socialization stuff" before going on the road? Nope. If you did that, you might end up spending your puppy's whole SPS at home. Socialization out in the world is vital during the SPS. COVID-19 may have made some things off-limits, but you can still safely introduce your puppy to a lot of the world outside his home. No longer is it recommended to keep your pup home until all his vaccinations are done. We are going to turn to the experts again so you can feel confident in your decisions about socializing your puppy.

The American Animal Hospital Association, in their Behavior Management Guidelines, states,

> *"If dogs and cats are deprived of appropriate exposure during critical sensitive periods, they have an increased risk of developing problematic behavior. . . . There is no medical reason to delay puppy and kitten classes or social exposure until the vaccination series is completed as long as exposure to sick animals is prohibited, basic hygiene is practiced, and diets are high quality. The risks attendant with missing social exposure far exceed any disease risk" (Hammerle et al., 2015).*

Additionally, the American Veterinary Society of Animal Behavior gives us an actual number. Their Position Statement on Puppy Socialization states,

> *"In general, puppies can start socialization classes as early as 7–8 weeks of age. Puppies should receive a minimum of one set of vaccines **at least seven days prior to the first class and a first deworming** [emphasis added]. They should be kept up-to-date on vaccines throughout the class" (American Veterinary Society of Animal Behavior, 2008a).*

We link the position statement at PuppySocialization.com/resources.

Again, the experts are emphasizing how important socialization is. So we follow the AVSAB's recommendation of waiting seven days after the first set of vaccines, then getting the pup carefully out into the world. If your pup got his first vaccines before coming home with you, be sure to verify the exact date so you can plan accordingly.

What You'll Need to Get Started

Socialization outside the home will involve the same learning principles you started using already: classical and operant conditioning.

To get started, you'll need the following:
- A hungry (not starving) puppy.
- Some kibble and some extra special, fabulous treats, as described in the chapter on socialization starting at home. Bring something in the form of small, soft bites that are very tasty. Small so your puppy doesn't get full, soft so he can eat the treat quickly, and extra special because you never know what you'll encounter out in the world. Novel situations need novel food, not everyday food. Our favorite higher

value food items are real meat (lean and cooked), cheese, and canned dog food in a squeeze tube.

- A stuffed Kong or something else long-lasting for the puppy to chew, depending on the outing.
- A plain buckle, snap, or martingale collar, or a non-restrictive shoulder harness. See PuppySocialization.com/resources for a linked article on choosing a harness. Don't use choke chains or prong collars, please.
- A regular, fixed-length leash (not a retractable leash).
- An interactive toy for you and your puppy to play with together. Our favorites are long toys appropriately sized for the puppy's mouth. Almost all puppies will engage in a gentle game of tug (gentle tug is good for puppies!).
- A familiar dog bed, blanket, or something else to put on the ground. When I (Marge) go out with a client's puppy, I usually bring a plastic platform or balance disc. Both of these are easily cleaned and sanitized. Remember to clean and/or launder any items that you used in public.
- A treat pouch (or wear clothing with pockets), so you have something to hold your treats and still have two hands free to manage your leash and your puppy.
- A bowl and water for your puppy.
- Waste bags for elimination cleanup. If you haven't done this before, it will seem gross at first. Don't worry; you'll get used to it. It's part of being a responsible pet owner. Put your hand into the waste bag and scoop up the waste with your covered hand (try to get it all). Then grab the open end of the bag and pull it over the waste and your hand. Now the waste is inside the bag.
- A crate or specially designed car harness. If you're taking your puppy someplace in a car, he should be safely confined and should not ride loose in your vehicle.
- Confidence. Wait. What? Where do you get that? Remember that you are in charge of these outings for your puppy. No other person, regardless of how well-meaning they are or what their station or position in life is, can be in charge. You are. You've done your research by getting this book. You have a plan. If you let someone derail your plan, there's a good chance you will regret it. We've both learned that from experience. But making a plan will give you confidence and sticking to it will give you even more.

You'll need the things above for all your outings, regardless of where you are going or what the goals of your outing are.

Keeping Track

Taking your puppy out into the world requires some planning. If you're using Dr. Yin's checklist, you should be looking for a variety of people, environments, animals, and objects to expose our puppy to. You want to ensure the outings are fun and include novel things, but aren't overwhelming.

It's hard to separate "socialization outings" into categories, because in the real world, everything is happening at the same time. You may have an "environments" outing, but chances are there will be people there. That's okay. There will be overlap. We're breaking it into categories to make sure you don't forget anything. And of course, if there is something too intense in the environment that you didn't plan for, say, people on a building roof using nail guns, pack it in or go somewhere else.

A great way to help you both plan *and* keep track of outings is to write out a list of different but specific locations. This could be a set of physical index cards or an app. I (Marge) did this when Zip was a puppy. Before I ever took Zip out, I sat down and thought of six or seven places to take him, then I created a separate card for each location. I made more as Zip progressed. Each time I went to that location, I noted on the card the date, time, and what the experience was like.

When I woke up in the morning, I didn't have to figure out where to take my puppy. I looked at my index cards and I could see my options. I don't know about you, but I have to plan ahead and write it down. The index cards worked for me to make sure I was including locations that would provide a variety of experiences and I didn't keep going back to the same two or three places. And recording the experience Zip had at each location allowed me to increase the intensity of exposures without leaping ahead or getting in a rut.

Riding in the Car

Socialization trips to new environments will most likely require a car ride. For your safety and your puppy's safety, he should be confined or restrained while in the car.

Remember to make the car a happy place for your puppy. You know how you can do that, right? Yes! By pairing it with food and play. Your puppy most likely rode home in a car to your house. How did he do? Was he quiet, fussy, or did he sleep? If he was calm and relaxed, that's great. If he was fussy, you have some extra work to do.

First, contact your veterinarian. Some puppies (and adult dogs) get carsick. Don't ask "Dr. Google" or your friends first. Contact your veterinarian. Your veterinarian is the leader of your puppy's care team and has been trained to care for your puppy's health. She may have tips or tricks to help your puppy ride better in the car. After examining your puppy and getting a history, she may prescribe medication to help your puppy ride more comfortably.

If car sickness is ruled out, and your pup is anxious or fearful in the car, your veterinarian may refer you to a behavior consultant or veterinary behaviorist. The process for helping your puppy will be the same one we have been describing and showing to you in this book. But when your pup is already fearful, it's good to have professional help. Being fearful in the car can have several possible triggers: the movement, the engine sounds, being confined—even the traffic sounds from outside the car might bother your puppy. A professional can help you address the problem most efficiently and successfully.

For a more confident puppy, or one who is just a little hesitant but generally bounces back, rides in the car can become classically conditioned quickly as positive experiences. This can happen almost automatically if you are taking him on fun socialization trips as we describe in this chapter. You'll be using the same process with the car that we describe throughout this book. Start easy and pair the experience with good stuff.

Make your first few rides in the car short and pleasant. Your puppy's only car rides should not be to the veterinarian. He could form negative associations with car rides if the only place he ever goes is to the vet. Take a long-lasting food item for your puppy to work on, like a stuffed Kong or chewy, and take a short ride with him—maybe around the block—then get out and play. If another family member can sit near him in the car, that can be a big help, too.

Car rides that include a yummy food toy while riding and that terminate in fun for the puppy generally create a dog who grows up to love riding in the car.

The next video shows a puppy riding quietly in a car while chewing on a food toy. I (Marge) was told this puppy didn't like to ride in the car. But she hopped right in and you can see her riding quietly while chewing on a food toy.

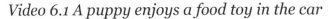

Video 6.1 A puppy enjoys a food toy in the car

In the adjacent photo, Zip is clearly comfortable riding in the car.

Planning Your Outings

We're almost there! It's time to plan where and when to take your puppy. Once he is comfortable in the car and it's seven days after he's had his first set of vaccinations, he can go!

Here are some final planning considerations before you take your puppy out. We want you to succeed by starting small (making things as easy as possible for your puppy), raising intensity when it's appropriate, optimizing your choice of location for success, and above all, observing your puppy and making sure his experience is positive. Here are some tips specific to observations, locations, and times of day that will help you achieve that goal.

We will be mentioning example locations for outings throughout the following sections, but will also have a more detailed list of possible locations at the end.

Start Small and Build

Within each environment you choose, there will likely be areas that are easier, with less novelty and stimulation, and areas that are more challenging. You will need to start small and build from there, only when your puppy is relaxed and happy at the current intensity.

So at first, start with locations and times of day **without** a lot of activity. The environments are already new to your puppy; you don't need a lot of strange new people, noises, dogs, and objects. You want to build on his experiences gradually by starting in quiet areas, without much going on, and gradually increasing the action or intensity as your puppy tells you he's comfortable.

A hierarchy of less intense to more intense will vary and be individual to your puppy. A home improvement store in a quiet, rural area will probably be less busy than a home improvement store in a suburban area. Use your best judgment when gauging the intensity levels of the available locations in your area. Choose less busy times for less intensity, busier times for more intensity.

Here's a sample progression of lower to higher intensity outdoor environments.

- A friend's house
- A quiet park
- The parking lot of an outdoor shopping mall at 8:30 a.m.
- The sidewalk of an outdoor shopping mall at 8:30 a.m.
- An outdoor café at an "off" time
- A sidewalk across the street from an elementary school during recess
- An outdoor shopping mall with wide sidewalks at 9:30 a.m.
- The parking lot of a hardware store at 10:00 a.m.
- An outdoor café at a slightly busier time
- A bench on a sidewalk outside a nursing home or senior center during the day

When I (Marge) first start going out with a puppy, I usually get up early. I want to get to the new environment while it is deserted. For example, if most of the stores open at 10:00 a.m., I will arrive at 8:30 or earlier. I go to the farthest end of the parking lot, away from the delivery trucks and people arriving early for work. As they do arrive, we observe them from a distance. I make sure we're not in the traffic pattern or hindering the businesses in any way.

Raising Intensity: Have a Conversation with Your Puppy

Whether you plan your socialization experiences or they present themselves suddenly as life is happening, make sure the intensity of the experience is appropriate for the puppy. Of course you can plan for the right intensity, but sometimes you will have to manage it on the fly. So plan to explore where and when you are most likely to find a calm environment for initial visits. A school parking lot will generally be less active on a Sunday. Meeting one or two small children at a park will be less intense than taking your puppy to a birthday party for small children. Sitting at a distance from a child's soccer game is less intense than sitting in bleachers when they are full. Plan for these favorable conditions, but be aware that any of these situations can change quickly.

As you go on more outings and get to know your puppy, you'll see that he might be ready for some exposures at higher intensity than others.

Before you build intensity while on an outing, we want you to ask your puppy two questions: **(1) "Can you eat?"** and **(2) "Can you play?"** By now, you are getting to know your puppy. Does he always eat offered food? Is he a "picky" eater? Is he taking and eating the food but with an extra "hard" mouth (ouch!)? Puppies can get shark-like and start grabbing for the food when excited or worried.

So pay attention to whether he can eat and *how* he is eating. If there is anything that tells you he is not relaxed and happy, this is not a good time to raise intensity.

Does he love to play with you at home but can't play with you in a new location? That usually means he is not comfortable enough in the new location. He may be scared or just distracted, but if he is not settling in, this is also not a good time to raise the intensity of an outing.

The answers to the questions you ask your puppy are valuable information to consider. You can use this information along with everything else you observe about your puppy's body language. Look at the whole puppy. Use what you observe to determine whether the current environmental intensity is appropriate for your puppy, too much, or if you can safely increase intensity.

You have to make these decisions all the time. Let's say you're at a quiet park and six children rush up and want to pet your puppy. You don't have young children living with you and this will be your puppy's first exposure to children. Here's where you may have to direct things. You manage the interaction to make sure your puppy doesn't get overwhelmed. Six children surrounding and petting a puppy all at once will likely be too intense an experience for a puppy who has not met children before. Consider it also an opportunity to educate children about dogs.

Make sure the kids first get permission from the parent or the adult with them before they interact with the puppy. Teach the kids that after asking their parent, they should ask the owner. If this is your puppy's first interaction with children, they should interact with the puppy one at a time. Six children surrounding and petting a puppy all at once will likely be too intense an experience for a puppy who has not met children before. Even one at a time, things could get intense for your puppy

The children should individually "ask" the puppy if the puppy wants to be petted by inviting the puppy to them. You can also teach them to pet "collar to tail," avoiding the head and legs. Be ready to keep the interactions brief. Depending on their ages and your ability to focus (prioritize your puppy!), you can explain why brief interactions are best to the kids, too.

If you are not comfortable doing any of this, or your puppy seems uninterested or nervous, you can politely and quickly leave. Try to get going before the children reach your space. Check out the later section, "If Your Puppy Gets Overwhelmed," for some tips. And yes, we are giving you permission to say "No." You are the expert on your puppy and know what is too much for him, and you are his advocate.

Picking Your Setup Area Strategically

It's also important to initially choose a spot where your puppy can experience and observe activity in one direction and not be surrounded by it. For example, if you go to a quiet park, choose a location within the park where most of the activity will be in front of you, not all around you. As you read your puppy's body language, you may increase intensity that same day or on later visits. Only increase intensity as your puppy shows you he is relaxed and happy and ready for more.

The next image is a simplified rendering of a park local to me (Marge). You can see the paved walking paths through the park, the tennis courts, and a playground. I typically start in the area marked "Low Intensity." There, the puppies are exposed to people and dogs only in front of them, with plenty of room to create distance from the path if needed. Notice there is also a stream between the walking path and that area—this creates physical separation. You can see me using the stream as a barrier in Video 2.1 (referenced on page 32).

The area marked "Higher Intensity" has plenty of room to create distance from the walking paths but is sandwiched between two paths. There can be activity on both paths at once. The area marked "Highest Intensity" is close to the tennis courts and a pavilion. There is less room to create distance and we are closer to action on both sides.

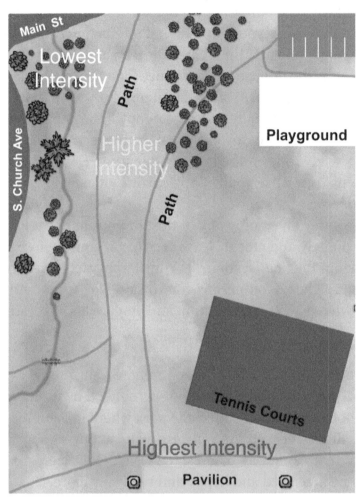

Locations, Locations, Locations!

Finally, before you go, here is some general information to consider about different possible locations to take your puppy.

- **Neighbors' yards.** Yes, your neighbor's yard is a completely new environment for your puppy! New yards can provide a way to let your puppy experience a new environment and people in a safe way. If your neighbor has a dog, be sure the dog is up-to-date on his vaccinations and worming. You'll need to be assertive about asking. Also, don't introduce the neighbor dog at the same time you visit their yard, even if you are planning an introduction at some point. One thing at a time!

- **Parks.** Look for parks with walkways (not dirt paths) and lots of space on either side of the walkway. I (Marge) choose parks that are likely to have people of all ages. I also look for parks where dogs are required to be leashed. Choose parks that have enough space to create distance, in case the puppy needs more space from something you encounter. It's important to remember, however, that parks do hold risks. Dogs of unknown vaccine history can frequent those areas, and you may occasionally encounter an off-leash dog. (That's one of the reasons you need space.) We know this is a lot to think about, but trust us, it's doable. And it's another good reason to start off in quiet, easy locations. You are learning, too!

- **Strip malls.** Choose locations in the mall based on the traffic patterns and your ability to increase distance from the action. For example, I've taken puppies to a small strip mall that had a parking lot in the back (make sure it's safe if you try this in your area!). If the puppy was comfortable there, then we'd enter a quiet courtyard. Then, if the puppy was still comfortable, we would walk by the stores.

- **Large parking areas.** Public parking areas during "off" hours can be quiet, novel areas. A library before hours or an empty school parking lot are both good places to check out. Always make sure walking your puppy in those areas is permitted and that it's safe for you to do so.

- **Hardware and farmers supply stores.** These large stores often have a lot of space. Be sure to check the store's rules first, as the rules about allowing dogs can vary. And remember that your goal is to socialize your puppy and not for everybody to coo over him. Keep your visit short and with some, but not a lot, of interaction with people.

- **Outdoor restaurants and cafés.** Outdoor restaurants and cafés that allow dogs can be good places to observe a variety of people and introduce new locations to your puppy. But be sure he has a lot of experience under his belt first. There may be people walking right up to or by him. To visit safely, take a mat, towel, or folded blanket with you. Set the puppy down on that with something to chew. Keep the first several outings short; maybe just order a beverage. And remember to go during the off hours. Your puppy is your primary focus on these outings. They are for him. If his body language and behavior tell you he is not comfortable in this environment, you should be prepared to leave. If it's an actual restaurant with service, be sure to tip well!

- **Types of locations you intend to frequent with your dog when he is an adult.** These might be family vacations, visiting grandchildren, agility trials, water camp, field work, or youth sporting events.
- **Competition and dog sport practice venues.** If you plan to do agility, obedience, or another sport, try to get your puppy in that environment. Sports clubs that require current vaccines for all participants are good places to help your pup get used to the noises, other dogs, and being crated in a noisy environment. Don't go there first, though! Wait until your puppy has gotten used to a lot of noise and activity in other venues, first. Be proactive about your puppy's safety. A good way to start at outdoor events is to create a home base at your car and make some forays closer to the action and back again, always judging your puppy's response.

Examples of Socialization Trips to Outdoor Environments

Take a look through the list of possible locations immediately above and think about what is available in your area. Whatever you choose, make sure you will have the ability to create distance, if necessary, from other people and activity for your first couple of trips.

We suggested asking your puppy, **"Can you eat?"** and **"Can you play?"** to get important information about his internal state on your outings. Throughout this section we will be suggesting more "questions" to ask your puppy. Things you can observe in his body language and little tests you can perform to check to see if he is doing okay. We hope these will become second nature to you. We will mention them explicitly in this section to give you an idea of how and when it's best to check in. If the answers indicate that he is worried or tentative, go to a quieter place, limit the novelty, or perhaps pack it in for the day.

For your first outings, choose areas that appear to be less frequented by people and dogs. If possible, park your car near where you will set up for your outing. That will limit your puppy's exposure in this public place. Leave your puppy safely restrained in the car while you take your chair, toys, poop bags, dog bed, water, food, and any other familiar items you brought along to your chosen location.

While you are setting up, take a final look around to make sure the area is as quiet as you have observed before. Do not hesitate to change your plans if there is too much going on. We want your puppy to have a fabulous first impression of going somewhere new.

When you've got everything set up the way you like, get your puppy. Leash him before you get him out of the car. Depending on the location you selected, you may elect to let your puppy walk to the area you set up or you may choose to carry him there.

As soon as he gets out of the vehicle, be ready. Give your puppy your full attention. Don't multitask by juggling equipment, texting, or talking on the phone. Observe and interact with your puppy. Once he is on the ground, note his body language. **Is his tail up, held neutrally, or down? Is his weight distributed equally or shifted back or forward? Does he look nervous or "relaxed and happy"?** These observations will help you help your puppy.

If all seems well with your pup, go right to the area you have set up. Put your puppy on the ground and observe him. What is he doing? What is his body language like? Is he frozen, immobile, and stiff? Is he sniffing and exploring? Is he moving toward the familiar objects? Can he eat food? Can he play with you? Will he chase a ball? Your puppy's answers to those questions will determine how you proceed.

If he is already interacting with things in the environment, such as looking at things, walking over to things, and sniffing things, you can already be using food. He walks up and puts his feet on a big rock: treat! Yes, we start that small. We want to start to fill up those bank accounts for the inevitable time when he bumps his nose against something or loses his footing. He gets treats for those incidents, too, or maybe his favorite toy, if he'll take it.

Some puppies just like to sit and observe, too. That's fine, as long as your puppy can take food (if he's not full). If he can't, move to a less busy location.

I (Marge) usually start with a quiet park. I choose one that has enough room for social distancing and for the puppy to be a good distance away from where people and dogs will be walking. I give him the opportunity to eliminate, get a drink of water, and explore the area. (The area should be a defined space—he doesn't need to explore the entire park.) I'm carefully observing him during this time as we describe above. If he is relaxed and happy, the next thing we do is play! I want puppies to think that going somewhere with their owners is going to be a fun adventure, not overwhelming.

The following video shows puppies playing on their first visit to a new location.

Video 6.2 Puppies play in new places

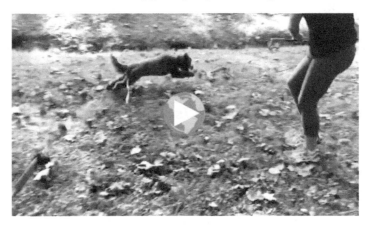

So let's say you've been in your quiet area for a little bit. How is your puppy doing? Keep asking questions. Can he look at you if you talk to him? Or does he need more time exploring the area? Is he ready to interact with you for a second or two? Can he do simple behaviors you taught at home (like hand target or sit)? Can he play tug or chase a ball or toy? The answers to those questions, combined with his body language, give you insight into how comfortable he is or isn't.

If your puppy is comfortable, you can move to an area of the park with more activity. It might be a playground area or a more populated path. At first, just move closer so your puppy can get a view of the activity; don't jump into the middle of a busy area. If your puppy is not comfortable, look for a way to reduce the intensity of the environment. You may choose to move further away or watch from your car. There's always the option of going home.

In the following video, you can see the puppy forming positive associations with children's voices and movements. She relaxed enough to lie down on her mat. Her owner does not have young children and wanted to make sure her dog was comfortable with them. She did a great job of planning ahead.

Video 6.3 Phoebe views active children from a comfortable distance

You can also interact with things in the environment. We want puppies to see novel items in the environment and be curious, not afraid. So look for ways for your puppy to interact with the environment. There's a bench. Perhaps your puppy can hop up on it (if that's safe), crawl under it, or walk around it? Remember to reinforce all those activities with food, play, or both. What about an empty swing set? See if your puppy can walk up to a stationary swing and investigate.

If your puppy happily explores the stationary swing, move him away from the swing a bit and then move the swing just a little (that increases the intensity) and immediately give your puppy a treat. **Did the swing make noise? What did your puppy's body language look like?** If he is relaxed and happy, you can add a bit more movement to the swing, or watch kids swinging (from a distance). If the movement worries him, keep the treats coming as you move him away from the swing area. You can work at home on things that move, like skateboards, balloons, and brooms, before trying a swing or other moving object out in the world. Save that for when he has more experience.

Your puppy doesn't have to do anything elaborate. Take a look at how the pup in the following video interacts with a bench. It's very simple, but as a result there is now money in the bank account of "benches are okay," and the credits might even start to extend to other stationary objects and parks in general.

Video 6.4 A mastiff puppy plays games with a bench

Puppies who fall in the "typical" category will generally explore the environment, eat food, play, and interact with familiar and novel objects. They will notice people and other dogs at a distance and not be too concerned. If your puppy shows signs of fear, see the chapter on fear and special challenges.

The following video shows a puppy's first socialization trip to an outdoor location in a suburban area. We used the time of day (early morning) to help reduce the intensity of the suburban shopping mall. Check out how loose and relaxed the puppy's movements are when he walks with his owner.

Video 6.5 A puppy plays and walks calmly around distractions with his owner at a mall

The puppy in the following video has been to this park several times. She's been in quiet parts of the park and also the more active parts, like the playground. She had a chance to observe cars at a distance, and she did not seem fearful. She was ready for a higher intensity location than a quiet park. In this video, she is playing in a pavilion near the road. She is being exposed to more car traffic, more foot traffic, and more noise overall.

Video 6.6 Phoebe plays with her owner near a lot of different activity

Remember how we said to start with small levels of intensity? Here's what "extra small" can look like. We've discussed Eileen's feral puppy Clara earlier in this book. Clara had to go much more slowly with meeting people than most. The following video of Eileen's puppy Clara shows what a short outing with the goal of exposing her to different substrates and environmental elements (without people) looked like. You can see Clara walking on a gravel driveway and on rocks close to a pond. Her body language is relaxed, she is interested in the environment, and she is able to play.

Video 6.7 Clara walks on gravel, grass, and near water

This next video shows another outing where Clara is watching a construction site from many yards away. She is a little cautious at first, but ends up relaxed enough to sprawl upside down and wallow in the grass. You can see Eileen giving her spray cheese when the construction gets noisy. It's hard to keep in mind that Clara was a feral puppy in most ways, because what you see in the videos is her complete acceptance of Eileen. But her space bubble for all other human beings before she tried to flee was generally 100 feet or more.

Video 6.8 Feral Clara views construction work from a safe distance

Clara had special needs. But even for a puppy with a more normal upbringing, you always need to have a plan and be ready to use your assertiveness skills to get your puppy "out of Dodge" if things get too intense. The next video shows Eileen letting Clara get a look at some activity in a grocery store parking lot from her car, and demonstrates Eileen's quick getaway plan.

Video 6.9 Clara sees a person in a parking lot and hears her car start

Did you notice that Eileen kept the spray cheese coming even after she mentioned that the woman they were watching went away? Listen and watch carefully; Clara subsequently focused on the noise of the car starting and its movement. Eileen gave treats for the noise and movement, too.

Things to Avoid

While we want your puppy to be socialized to a variety of things, do your best to avoid the following.

Dog parks. Avoid dog parks with your puppy. You don't want to take your puppy to places unvaccinated dogs might frequent. Besides the risk from feces, some owners bring aggressive dogs to dog parks and certain combinations of dogs are likely to be overwhelming for your puppy. And despite their popularity, a lot of inappropriate dog play gets practiced at dog parks. Don't risk it.

Dogs with unknown vaccination histories or dogs not up-to-date on their vaccinations. Be sure that all dogs your puppy meets are up-to-date on their vaccinations. Yes, this means you need to ask your friends and family if their dogs are current on their vaccinations and healthy.

Animal feces. Watch out particularly in parks, but poop can be anywhere. You need to make sure that your puppy doesn't come in contact with any.

Dogs who might not be good with puppies. You need to be super careful about your puppy meeting unknown, adult dogs. Not all dogs are good with puppies. Puppies, like babies, don't understand social norms or signaling. They might invade a dog's space, put body parts in their mouth (with those sharp little teeth), climb on the dog, and do things that hurt! Some dogs are wonderfully tolerant with puppies, but others may be masters with adult dogs, but not tolerant of puppies. This is an important difference.

I (Marge) learned this the hard way. I thought I'd very carefully chosen an adult dog with a rock solid temperament for Zip's first dog introduction. I'd seen this dog defuse inappropriate interactions with other dogs several times. Zip ran up and put his paws on the dog and received an over-the-top correction that left him whimpering. Oops! The dog who was masterful at defusing conflict with grown dogs was *not* tolerant of puppies. Or maybe he used to be, but now was older and in pain. The point is: if you don't know the dog is good with puppies, don't take the risk. It's not worth it.

<u>Toddlers.</u> Puppies and toddlers: talk about a combustible combination! Toddlers make jerky movements and don't have impulse control. Parents may not be fast enough to intervene, and sometimes don't understand the need to do so. A toddler and a puppy can scare each other badly, or one can even get hurt. Toddlers are great for your puppy to watch from a distance, but not to interact with.

We know this is a lot to think about, but trust us, it's doable. And it's another good reason to start off in quiet, easy locations. You are learning, too!

What If You Go Out for Socialization and Nothing Happens?

We tend to get super goal-oriented when we are socializing our puppies. It's natural; there are so many things we want our puppy to learn about. So what if we plan an outing, gather our equipment, get the puppy, drive to a location—and nothing happens? Where is that dog who usually gets walked in the distance? Why isn't anyone at the outdoor café? They don't even have their speakers blaring!

Don't let that frustrate you. Your puppy practiced riding in the car and getting out with you in a new, quiet location. That's fantastic! Play some training games or give your pup a food toy and just hang out. You can both relax and watch the world go by. Puppies need time to learn to relax in new environments, too. We don't want to bombard your puppy every time he goes out. "Relax and chill" is an important and valuable skill. We want you to make "relax and chill" in a new environment a priority, too. Congratulations on your outing where "nothing happened." You helped your puppy form positive associations in a new place and that is part of socialization, too.

Meeting People Out in the World

When you're ready to expose your puppy to people outside your home, you don't have to start with actually *meeting* people. We recommend you start with *watching* people.

When Zip was little, I (Marge) took him to a sports field to watch children play sports. I don't have children, so I went out of my way to help him form positive associations with little humans. We hung out several feet away from a sidewalk. We safely sat on a blanket I brought along and watched and listened to the kids. You can see the moment in the next photo.

What you don't see is that I was behind him with a toy and a pocketful of food. I was ready to run interference if someone decided to approach him, but I picked a place where that is less likely to happen. The kids on the sidewalk were hurrying to a sports field and were not likely to stop to see the puppy. From that vantage point, he could also see different silhouettes (shoulder pads, helmets, people carrying things, people squatting) and was exposed to a variety of different people.

We also sat on a bench near a nursing home. He formed positive associations with people using walkers and wheelchairs.

Those are just a couple of the places we went. I bet you can think of more for your puppy. Remember, too, that exposure alone is not socialization. Forming positive associations with those exposures is the goal. On the other hand, your puppy doesn't have to interact with all those people. Having a good time (food and play, remember?) in their presence is often enough.

But let's say you have watched a lot of people with your puppy and you're ready to do some actual greeting (and it's safe to do so—this might not happen in places where the COVID-19 pandemic is at severe levels).

First, review our guidelines for greetings: "Puppies Meeting People" starting on page 119. We hope you have been practicing them at home, but here's a quick review of the one we think will help you the most on the road: the Three-Second Rule.

The Three-Second Rule (Review)

We described the Three-Second Rule in detail in chapter 5. You probably needed it at home, but you definitely need it now, if you introduce your puppy to strangers out in the world.

If a stranger acts interested in your puppy or asks to visit with him, and you think it will work out, use these steps to manage the interaction:

1. Let the person know it's the puppy's choice to greet or not, and tell them up front the puppy may be tired. This last remark gives you a chance to bail politely if need be.

2. Let the person know you are training your puppy. Say that you'll be calling your puppy back to you from time to time so he gets practice listening to you when he's doing something else.

3. Tell the person that when you call the puppy, they should immediately stop interacting with the puppy.

4. Here's the Three-Second Rule part: Allow your puppy to greet for three seconds (you can count "one banana, two banana, three banana" to yourself) while you or the stranger feed him treats. Then call the puppy back to you and give him a treat.

Remember to observe the greeter and your puppy's body language throughout the interaction. If your puppy looks comfortable with the interaction and the greeter is appropriate and doesn't scare your puppy, you can release the puppy to go interact with the person again.

Socialization with People Gone Sadly Wrong

We both have many stories about this.

I (Marge) once saw someone interact with a scared puppy while at dinner. The puppy was with his owners at the outdoor seating area of a restaurant. The couple was approached by a dog lover who asked if he could pet their puppy. The owners said "yes" and explained that the puppy had washed out of a service dog program for being timid. The tall stranger approached the puppy and the puppy yawned, lip licked, and I could see "half-moon eye" from a table away. The puppy was stiff and still. The stranger had every good intention of showing affection to the puppy, and stroked the puppy with one hand along his back. He then put his face in the puppy's face and kissed and petted him. The puppy froze with the whites of his eyes showing even more. He was terrified. The person petting him had no idea he had frightened the puppy. He wanted to pet and interact with the puppy, did so, and was quite satisfied.

The puppy's owners also had no idea their puppy was terrified. They probably thought they were "socializing" him. It's a good thing the stranger didn't get bitten. Everyone was fortunate the puppy didn't bite. He used the nicest language he knew to say he was uncomfortable with the contact from the stranger, and it wasn't enough.

Will that puppy be more worried or less worried the next time a stranger approaches? He's going to be more worried. That experience did nothing to make the puppy more comfortable with strangers. Instead, it showed him that people will disregard him when he tries to say he's uncomfortable with the contact.

Please keep this story in mind when strangers interact with your puppy. Ask yourself a few questions. Is the experience with this person making your puppy feel better/safer interacting with people? Is it neutral? Will it make him feel worse? Neutral experiences are not enough. Remember the bank accounts? Your job is to keep making plenty of deposits in the "unknown people are good" bank account. Neutral encounters don't add deposits, and experiences like the one I witnessed make withdrawals. Make sure your puppy's accounts are full of good experiences.

Eileen now. Here's my "mini-disaster" story. It's a cautionary tale about the unpredictability of humans. I was trying to do everything right.

When I was working on exposing Clara to the human world (we won't call it socialization because it was way outside her SPS), we often went with my trainer to a certain shopping center. I was acquainted with a clerk in a store there, and she was very curious about Clara. At this point, it was rare for my trainer and me to allow any stranger to get close to Clara, much less give her a treat. But I talked to my trainer and we figured it would be worth a try if we were careful. It was a rare opportunity to provide a quick, controlled visit with a stranger, or so I thought.

So I talked and talked and talked to my friend about how Clara was different: don't look at Clara; don't hang around after the treat; do this and this, but not that. So the day arrived when I was ready to let my friend greet Clara. Clara and I got there and were waiting outside the store. My friend was a little late because she had to wait for her break. Then the door banged open and she burst out, yelling, "HI!! HERE I AM!!" and ran straight at us. Best laid plans.

That story highlights one more difficulty. Giving specific, direct instructions to strangers or people you don't know well can be awkward. Here's this nice person who offered to help, and you are giving them orders! Or worse, retreating away from them when you had asked them to come help you! Just remember that you are your pup's advocate. Keep your eyes on the prize, which is to have a happy and confident dog. Sometimes you'll have to do as I did: get the heck out of there and apologize later.

I call this event a mini-disaster, not a total disaster. This is because even though Clara's default setting was to be afraid of people, we gave her so many careful, positive exposures that this encounter ended up being only a blip on the screen. But I was lucky. You can never predict the long-term consequences of scaring your dog.

Your Puppy Doesn't Need to Love Every Human or Every Dog

Do you love every human you meet? Not likely. Dogs have preferences, just as people do. The goal of socialization is for your puppy to become comfortable in the human world. Some dogs really do like everybody. They just like people. Others like *some* people. We want to make sure they aren't making choices based on lack of exposure. Maybe your pup would have liked Uncle Bill if he had met some other extra-tall men with clompy boots when he was a puppy. Or if he had been exposed to enough men of all sorts he could put Uncle Bill in the category of "Men—safe." We never recommend any contact with humans that the pup isn't comfortable with. Surprisingly, this lack of force is a good way to get them comfortable with people. It helps puppies if we allow them to explore at their own rate, while providing support with yummy food and fun.

Again, watching body language is key. If your puppy shrinks away from Uncle Bill who is sitting there quietly, it's time to do some work. If he gives him a sniff and wanders away to check out something else, that's probably fine. Maybe he'll end up being pals with Uncle Bill or maybe not, but his decision won't be based on fear.

Likewise, your puppy may not end up being special friends with other dogs, even others in your home. We all hope for him to have a buddy, of course, but some dogs are fairly indifferent to other dogs, even ones they live with. We want your puppy to be comfortable with and unafraid of the other safe dogs he lives with. If they become special pals, that's a bonus.

Planning Ahead for Indoor Environments

By indoor environments we mean stores that allow pets on leash. Before you undertake a trip to an indoor environment, call the business and make sure pets are allowed if you are the least bit unsure. (Some individual stores that are part of national chains have regulations that are in conflict with the website information.) It's also a good idea to go on your own and scope it out, even if you have been there many times before. If your previous trips didn't include a puppy, you may not have noticed something that could make or break your trip. So take another look with your "puppy socialization" eyes.

Indoor environments can be trickier for many puppies because they are more confining, with less room to escape. It is easier for puppies to become trapped, so keep that in mind as you start exposing your puppy to new indoor places. This is another good reason to go ahead of time and look at the layout of the store and property. You can map escape routes.

Also keep in mind that you are more likely to encounter people who will want to interact with your puppy. Will you allow it? If yes, how will you manage the interaction? If not, what will you say? Make a plan and decide ahead of time how to manage or discourage interactions. Remember: it's a human tendency to want to show the world your puppy, but that is not the same as showing your puppy the world! Resist the temptation to let your puppy become a magnet for human attention. The last thing you want to do is let him get overwhelmed and frightened.

Example of a Socialization Trip to an Indoor Environment

In the following video, I (Marge) am meeting a client at a tractor supply store for training. We don't start inside; we start outside, in a quiet corner of the parking lot. We give the puppy the chance to eliminate and observe her new location. When she can play and respond to simple cues, we move closer to the covered area near the store. There are new things to learn about there (shopping carts!). We observe the sliding doors and people coming and going at a distance before we finally move indoors for the last 10–15 minutes.

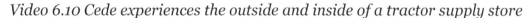

Video 6.10 Cede experiences the outside and inside of a tractor supply store

What can we learn from the outing to an indoor location with the golden retriever puppy in the previous video? **Going indoors starts outdoors.** Remember that there may be challenges before you even get into the building. Arriving at a location without a plan and marching inside with a puppy is not a great idea, and exactly what we *don't* recommend in this book. So here's what to do.

Always take food and a toy with you, even to indoor locations. As a courtesy to others, make sure the food you take will not shred or leave a mess as the puppy eats it. It's good to keep a small footprint when in public and make sure to potty and clean up after your puppy before you go inside and to give him frequent breaks. Also be careful not to interfere with others who are shopping or working.

Get out of the car yourself and scope things out first as we described above for outdoor locations. Some stores have a lot of activity in the parking lot so take a good look around. Put your puppy down in a quiet part of the parking lot and observe your puppy "that day." By this we mean to drop any assumptions like, "He had an easy day yesterday so he should be up for this" or, "He's been here before, so I can start out with more challenges." No. It doesn't matter if he's been here or to a million other locations on other days. Observe his body language that day. Is his tail high, low, or neutral? Is his weight distributed equally? Is he excited to move forward and explore or is he leaning away from the action?

Before you can get inside to many places that allow dogs, you often have to get by those darn automatic sliding doors. Their sudden movement and noise can startle a puppy! I (Marge) let puppies observe the doors opening and closing from a distance first. And of course, when the doors move at a distance, I give the puppy treats, so the movement predicts treats.

After you've allowed your puppy to acclimate outside and get used to the doors without showing any fear, you can go through the doors.

Once you're inside, make a beeline for the quietest part of the store. You don't want your baby puppy to walk into a new location and be swarmed by a bunch of strangers or overwhelmed by the season's giant inflatable holiday decorations. Find a part of the store that's not as busy and focus on and interact with your puppy. Settle down and make sure your puppy can still take food and play. Continue to observe, check, and recheck his body language.

Keep the first visit easy and quiet. If the location turns out to be a good one where your puppy gets comfortable, you can probably meet and interact with some people there on a later visit. But this first time, dodge the other humans as much as you can and keep close attention to how your pup is doing.

Objects Outside the Home

Generally, approach objects out in the world using the same guidelines we described for household objects in chapter 5. But objects in the real world are not as predictable. They may be big or seem extra-weird to your puppy. They may blow around, move unexpectedly, or they may make noise—all in ways that aren't under your control.

You can help your puppy by planning interactions with lower intensity to higher intensity items away from home. Your list might look like this:
- A big rock (something your puppy can easily climb on)
- A bench (your puppy can go around, under, or put two paws on it)
- A stationary swing
- A doorway
- A moving swing
- Playground items (first stationary, then moving)

When I (Marge) go out with a puppy, I try to make the environment our playground. I look for stationary things to interact with: rocks, logs, stumps, garbage cans, and park benches. I also look for things that move: balloons, kids on bikes, shopping carts, automatic doors, and elevators. I want my puppy to have the opportunity to observe or interact with things that are part of the human world, things that might not normally be on my radar because they are commonplace. Within the boundaries of courtesy and reason, I want my puppy to think the world is his playground. I pair each interaction with food or play (or both!) and plenty of social interaction from me.

The puppy in the adjacent photo is investigating a statue at our local park. Statues of people and oversized animals can be super-weird to dogs, so we were lucky to have this one to interact with.

The following photo shows a client and her puppy having a fun time playing with tree stumps at a park.

In the following video a puppy investigates a shopping cart. You can see that we start with it stationary, then we add some movement. At the moment the puppy stopped eating treats, we should have slowed way down. Typical, non-fearful puppies are resilient in the SPS. This puppy recovered nicely. If I could have a "do-over," I'd take a bit more time with the stationary cart and let the puppy approach the cart again on his own after the movement.

Video 6.11 Mason learns about shopping carts and their movement

The puppy in the following video came home during the "shelter-in-place" requirements of the COVID-19 pandemic. She did not experience locations other than her home. When she first encountered this playground equipment, she was frightened. Her owner did a fantastic job of allowing her to progress at her own pace instead of trying to lure her toward the scary piece of equipment.

Video 6.12 A puppy learns that she doesn't need to be scared of this piece of playground equipment

Don't forget to expose and socialize your dog to holiday decorations when you have the chance. The adjacent photo shows Zip with some Halloween decorations, then you can check out a video of another puppy investigating a different type of decoration.

Video 6.13 Cede meets an Easter decoration

As we've said, Eileen's puppy Clara had to be introduced very slowly to people, but Eileen helped her build positive associations to a lot of moving objects when she was young. As an adult, she finds these over-the-top Halloween decorations with flapping dragons no big deal. They included weird noises, too, from their loud, motorized fans. You can see her comfortable body language in the following video.

Video 6.14 Clara is confident around some huge, noisy Halloween decorations

Sounds Outside the Home

Just as you do at home, make sure you're ready with treats and toys in case there is a sudden noise when you are out with your puppy. What if a transformer blows with a loud pop, an airplane flies over, or your neighbor's screen door slams? Make it "treat city" for your puppy.

I (Eileen) continue to give treats when loud noises occur throughout my dogs' lives because it is natural to be scared by sudden sounds. It would be tempting to say that taking proactive efforts could prevent sound sensitivity in the future, but we don't know if that's true. Some scientists believe sound phobias in dogs have a genetic component (Tiira et al., 2016). But it does seem like a good bank account to keep topped up.

Zip was introduced to a variety of noises at an early age. He's 13 months old in the following video, and this wood chipper isn't even a distraction.

Video 6.15 Zip can easily work near the sound of the wood chipper

Sounds can definitely be challenging. Sometimes they appear to come from nowhere like thunder or some electronic beeps. Others have a clear source, like the noise of a vacuum cleaner or lawnmower. In either case, pair food and play with the sound. In the following video, the puppy engages in play with noises at a distance.

Video 6.16 Tinker happily plays while there is yard equipment noise in the background

So for unplanned, sudden sounds, be ready with treats and toys. But what if you want to expose your puppy in a controlled way to sounds that might be scary?

Remember the process we described with the vacuum cleaner? We described getting the pup used to it from farther away and with short exposures, then getting closer to increase the intensity. All the activity was paired with great treats, of course.

But think about how much control you had over that situation. Keep in mind that a lot of noisy things outside are not safe for dogs to approach. Getting close is not the goal. You don't want your dog to fall in love with the lawnmower or electric hedge clippers. Okay, so we do know one dog who really enjoyed a leaf blower! We linked the cute video at PuppySocialization.com/resources. You just want them to be comfortable with the sound at a safe distance.

Also, many noisy things outside aren't under your control. They are unpredictable. You always need to be ready to end the session quickly if your puppy isn't handling it. For example, let's say you are in your driveway with your puppy. You have roast chicken with you, a very high value treat (because you never know what will happen when you are outside).

You hear a garbage truck five houses down. Your puppy notices the truck and the workers, and you immediately start feeding small pieces of chicken one after another. Your puppy is watching and eating. Then the truck moves closer and now it's four houses away. Your puppy is watching and eating continuously. Then the truck moves again and it's three houses away and your puppy abruptly stops eating. Your puppy has gone into a state of fight, flight, or freeze. His body won't let him eat. You're too close to what's worrying your puppy and he's no longer forming positive associations with garbage trucks. Make a note of the conditions, such as how close the truck was when pup stopped eating, and head inside. Get him out of the situation. And most important: next time, go inside before your puppy might hit "fight, flight, or freeze."

Sometimes sounds go on a whole lot longer than we expect. The next video shows Marge dealing with a garbage truck that stayed much longer than usual. I (Eileen) am going to describe some highlights.

Marge had a board and train puppy named Irving whom she knew to be sensitive to loud noises. At the beginning of the week, he would fear-bolt to the house when airplanes went overhead and if loud, noisy trucks went by. The first day he was with her, he heard a loud truck about three blocks away and bolted toward the house. But with careful work by Marge, within a week he could sit and watch a garbage truck at the end of the driveway.

Marge, as always for such an outing, had very high value food and a toy with her. And it's a good thing, because after loading the garbage, the workers stayed there and compressed the load right at the end of her driveway.

Note in the video how Marge keeps the food coming. Irving stopped being able to play (6 seconds into the video), but he could eat. The minute and a half the truck was in front of the driveway takes forever to pass, doesn't it! Watching carefully, I can see

Marge deliver at least 20 treats during that time. And she never stops talking cheerily to Irving.

Take a look at the video and watch how generous Marge is with the food. She keeps it coming for the duration of the presence of the garbage truck. Using great food and lots of it is what got Irving to this point, and it's what kept him able to stay in the presence of the truck.

Video 6.17 Irving stays in the presence of a persistent garbage truck with help from Marge

One more thing: Marge is a professional dog trainer and an expert at reading a dog's comfort level. She assessed that Irving would be okay, with some help, with this long exposure. But if I or most of you were in that situation, it would be a better choice to be prudent and get the pup back in the house, as Marge and I recommended earlier.

In an ideal world, we'd always stop before the puppy hits fight, flight, or freeze. Those instances can set our training back. But sometimes we don't know where this ever-moving spot is until we've gone too far. If that happens, increase distance from the scary thing as soon as you do notice.

It would be great if we could control exposure to all sorts of potentially spooky things out in the world. But we don't always have control over these things. You can be ready, however, with treats and toys, and the ability to read your puppy's body language, to help him be comfortable around sounds.

Dogs from Outside the Household

The best way to start introducing your puppy to other animals of his own kind is to introduce him to other healthy puppies who are also up-to-date on their vaccinations. That's part of the reason puppy classes are so popular. We discuss puppy classes in a later section; right now let's talk about introducing your puppy to another pup or adult dog in a less formal setting.

We know this activity is harder and in some cases impossible because of the COVID-19 pandemic. But if you are able to let your puppy visit with other puppies within your safe social bubble, we hope you'll do so. If you arrange any outdoor puppy get-togethers, be sure to follow all local ordinances and current safety precautions. We link to a page from the Centers of Disease Control and Prevention (CDC) covering COVID-19 and pets at PuppySocialization.com/resources.

If you can safely do so, make a special effort to get together with other puppy owners. Besides taking mutual precautions about COVID-19 exposure, make sure they are following the standard guidelines for their pups' vaccinations.

Playing with other puppies is an important part of your puppy's behavioral development. He learns bite inhibition (yay!) along with play and social skills. But puppy play should not be a free-for-all. Owners should monitor play closely to make sure all puppies are enjoying the play. Play styles vary between breeds and one puppy's play style might be too exuberant or physical for another puppy. Remember, your puppy is learning what interacting with other puppies and dogs is like. Make sure he's learning the lessons you want him to.

How do you know if your puppy is enjoying the interactions? Well, you're ahead of most of the other owners because you've studied body language. Watch closely! If it seems like one puppy is always on top or one is always being chased, separate the puppies for several seconds. Release the puppy who's been on the bottom (or been chased) first. Does he re-engage the other puppy? Or does he look for something else to do? Let his behavior be your guide. If he chooses another activity, don't let the puppy who was being pushy bother him. You have to pay close attention to your puppy's interactions with other puppies 100% of the time at the beginning. It's not the time to chat with friends, distracted, while the puppies play. You must watch to make sure your puppy is enjoying the interactions and that the puppies don't need a break.

Marge here. Many of my clients have neighbors or family members with friendly dogs. Without much effort on their part, they have a pre-established social group for their puppy. They are fortunate! And if they're even luckier, they may even have a friend who has a puppy at the same time.

If you don't have that, make an effort to find some puppy-friendly dogs to meet your puppy. I remind my clients who want to set up meet and greets to make sure they talk with the owners, especially of grown dogs, to make sure the dogs are good with puppies. Not all dogs are.

Ask your friends and family members whether their dogs played with other dogs as puppies or already have other canine playmates. Those are good indicators to tell you whether the dogs "play well with others." Avoid older dogs whose owners say they "need socialization." What that usually indicates is they don't understand the details of what socialization means, and that their dog may not be appropriate around puppies or other dogs. And just as you did with any resident dogs, take it slow. I recommend starting with a parallel walk, walking side-by-side with some space between the puppy and other dog. As you do this, observe the dog's and your puppy's body language. Does it look loose and soft? Or stiff and hard? Listen to and advocate for your puppy. Move your puppy away from the adult dog if either one of them exhibits body language that is stiff or in some other way uncomfortable. End the session if your puppy is not forming positive associations.

I interrupt play often between puppies and between puppies and adult dogs. Dogs naturally take little micro-breaks in their play. Puppies have to learn that skill. If they play continuously without any breaks or interruptions, they can get overexcited. Before you know it, a fight breaks out.

Zip has done a good job with puppies I've brought into our home. In the following video, you can see Zip with Cocoa, another Portuguese water dog. Zip teaches Cocoa his "dead bug" move. What I like about these interactions is the soft, loose body language of both dogs when they are indoors and outdoors. What else do you see?

Video 6.18 Zip engages in friendly play with Cocoa

Other Animals

You will likely need to socialize your puppy to some other animals. You might be thinking, "I don't live on a farm. Why do I have to worry about other animals?" Cats, ducks, geese, horses, cows, squirrels—your puppy may encounter many of these during his lifetime, even if you don't live on a farm.

I (Marge) often had migrating ducks and geese at the ponds in my neighborhood. One duck would follow us on our heeling pattern when we trained at the park! In Florida, agility trials were often held at an equestrian center. It was not unusual to see horses on the grounds. At my current home, there are cows next door. A lot of people keep chickens. All these creatures do something most dogs find exciting, including that most retreat rapidly. Ah, potential chase objects! Except geese, swans, and even some ducks are just as likely to advance and aggress if provoked. Take care with any birds or animals you watch!

I typically expose puppies to other animals in a controlled way. That means on leash. I don't make a big deal about the other animal—no direct introductions. I'd like my puppy to be aware, but not too curious, fascinated, or over-aroused. There is no reason my dog needs to be friends with a duck or cow, unless they are part of my animal family. We don't need to get close. When the dog or puppy notices the animal, I automatically feed and interact with him. I want him to form positive associations with the sight of other animals and not be afraid. I use the leash to make sure my puppy doesn't frighten or chase the animal.

The team in the photo below is observing waterfowl from a safe distance.

Your dog or puppy may become afraid or overexcited by the sight of a cow or squirrel. If that happens, increase distance from whatever is the cause of the fear or excitement. You want to find the point where your puppy is aware of the animal, but not so locked in on it that he can't interact with you. And that's the location where you observe and use food and play. You can move closer if your puppy gets comfortable, but keep in mind that your goals with such animals are different from a lot of the things we have talked about. You aren't preparing your puppy to interact with them, just to feel comfortable when seeing them.

Of course, your choices may be different if you plan to use your dog for herding or farmwork. And what happens if your puppy does become obsessed with squirrels or chipmunks? That's usually more of a training issue than one of socialization and is beyond the scope of this book. A good trainer or behavior consultant should be able to help.

The Veterinary Clinic

These sections on the veterinary clinic and the groomer could have gone earlier in the book since vets and groomers are people (really!). However, we chose to put them here after you'd already learned a lot about socializing your dog to new people. These professionals work in specialized environments/locations. They will often cause weird noises to happen. For many puppies, the veterinary clinic and grooming salon are extra scary places. Unpleasant things can happen there. As your puppy learns about his world, don't forget to help him form positive associations at the veterinary clinic.

You need to counter the negative associations that are a natural response to early veterinary care: those shots, sprays, and temperature checks. Remember bank accounts? The veterinary clinic starts making withdrawals from their bank account right away, often before there have been any deposits. So, make sure you make deposits regularly throughout and even before those visits.

We recommend that you take your puppy for "happy visits"—visits without any procedures done. During the happy visit, give your puppy yummy treats and perhaps have some fun interactions: this puts deposits in the bank account. Almost all the veterinarians we've talked with want their puppy owners to bring their puppies for happy visits. Contact your veterinarian today to find out what their happy visit policy is. Even during the COVID-19 pandemic, there may be something they can offer, even if it's a tech coming out to the parking lot with an otoscope or stethoscope. But plan to pay for a visit if they can offer you any time at all. Most veterinarians are impossibly busy right now.

If your veterinary clinic can't currently provide happy visits, even car rides to the parking lot that are paired with treat and play parties (in the car) can help. Pick a nice day and roll your windows down when you get there. If you've had dogs before, you've probably noticed that dogs learn to recognize the look and smell of that parking lot and sometimes even learn the driving route to the vet. Any positive associations you can add will help.

The Groomer

If your puppy will have to be groomed, start practicing early. Marge here. When Zip was still with his breeder, she introduced the puppies to the grooming table, brushing, and clippers. Even so, I spent time at home pairing the noise and vibration of the clippers with food.

When it was time to groom Zip for the first time I spread peanut butter on the arm of the grooming table. In this video you can see it keeps my wiggly 8-week-old puppy busy while I groom him.

Video 6.19 Young Zip licks yummy peanut butter in an early grooming session

When Zip got faster at licking the peanut butter off the arm, I stuffed his puppy food toy with a favorite food (his was salmon) and tied it to the arm of the grooming table. You can see the setup in the adjacent photo. He'd stand and eat while I brushed or clipped him. Now, he eagerly jumps up on the table when the clippers or brush come out. The early association with yummy treats paid off bigtime.

Zip became very relaxed on the table and with the whole grooming process.

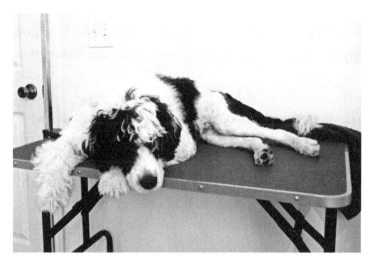

If you have a breed that needs grooming and you don't plan on doing it yourself, be sure to get your puppy used to brushing (see "Handling" in chapter 5) and the groomer before he needs his first bath or groom. Start at home so he can form positive associations with that activity in a comfortable environment. The grooming salon can be a scary place for dogs who were not exposed to it as puppies. Grooming salons often keep dogs in crates or kennels while they are waiting for their bath or groom. If your dog isn't used to a crate, you should start crate training him so it won't cause him stress. Some groomers use "cage dryers" with air blown into the cage to help dry your dog. At the grooming salon, there are other dogs, clipper noises, dryer noises, and more. You can see this could be frightening for a dog who did not form positive experiences with those things as a puppy.

Do yourself a favor: help your puppy form positive associations with those things when he is young. Schedule some short happy visits with your groomer at a time when the salon is quiet and not overwhelming. Take yummy treats with you so you or the groomer can give them to your puppy, and keep those early visits short and sweet. Pay your groomer for these visits, or if she won't accept payment, send her a little gift.

If Your Puppy Gets Overwhelmed: Planning for the Unexpected

Even puppies who are not abnormally fearful have hard days sometimes. Your puppy might be over-excited, worried, or just tired. Or, your puppy may be doing fine, but then something unexpected happens at a higher intensity than your puppy is ready for. For example, let's say he's seen cars at a distance and shown no reaction. Now you're walking down a quiet street and a car suddenly zooms by and he tries to bolt. What can you do? Help him. In this section we'll show you some clips of actual "suddenly in over your head" situations, then list some strategies at the end of this section.

Here is a real-life example of being in over the puppy's head. My client and I (Marge) were too close to the children and activity in this clip. Even though the puppy is eating food, he does not look "relaxed and happy." He looks worriedly at the children when they make "kid noises." His ears are to the side, his tail is low, and his back rounded. He starts to bark. We were preparing to move during this clip, but paused when he offered a sit. It was great that he had enough composure to sit, but it became clear that he was still too uncomfortable to stay. We moved further away immediately.

Video 6.20 This puppy becomes upset while he is a little too close to noisy children

In the next clip, puppy Cocoa is worried by the traffic. (Yes, in case you were scratching your head, we do have two different puppies named Cocoa in this book.) As we walk along a side road, we are too close to cars passing on the next road. Cocoa puts on the brakes and doesn't want to move any closer. When she starts walking again, she lags behind her owner, reluctant to move forward. She crouches down in response to a rumbling traffic noise. I then stopped recording to coach the owners.

Here's the video up to that point. Watch with the sound off first. The happy talk from the humans can influence your perception of the puppy.

Video 6.21 Puppy Cocoa shows some subtle signs of unease

This clip shows the more subtle behavior you'll need to learn to watch for. Cocoa was not whining or trying to bolt. After walking tentatively, she sat down. Well okay, puppies do sit. But relaxed and happy puppies tend to keep moving on a walk. Then, she not only lagged, but at the very end, just before I had to turn off the camera, she crouched and dropped her head at a louder noise. If you were out with your puppy and looking at your phone, or even had your head turned to talk to a companion when the pup did this, you could have missed most of it. The result: you would force your puppy toward something that scared him.

But we think that by now you have built up some great skills in observing your puppy. You can catch the more subtle behaviors that tell you he is getting anxious or unhappy with the situation. But whether you catch the problem early or later, generally the best thing to do is to increase the distance between the puppy and the scary thing.

As soon as Cocoa got scared of the traffic, we increased the distance from the traffic to a point where Cocoa could play (she loves to play). I took the next video just a few minutes later. See that bounce-back? Cocoa is out of her SPS at this point, but because we noticed and acted as soon as we saw she was scared, she recovers quickly.

Video 6.22 Cocoa happily plays when farther from the cars

These unexpected events are going to happen. Here are some things you can do:

1. Shorten your leash. Shorten your leash so you have more control. Shortening your leash means sliding one of your hands closer to the collar or harness. The more leash there is between your hands and the dog, the less control you have when something happens. It doesn't mean adding pressure and holding the dog back, because that usually makes things worse!

2. Use movement and increase distance. Turn around, make sure you're facing the way you want to go, and start moving. In a happy voice, start talking to your puppy. In almost all cases, your puppy will come along. He may hesitate, but he'll likely come with you. Be careful not to jerk on your puppy or pull him off his feet. You can also make sure your body says, "We're leaving now," which we describe below.

Many people try to call their puppy away from the distraction or scary thing, but their body is saying the opposite and the way they call doesn't help. They face the back end of the puppy while he's looking at the distraction and they call the puppy's name. First of all, your little baby puppy does not have the skill to be called away from a distraction at this age. Second, if he should happen to glance at you, your body language is telling him to stay back, because you're looking at him and the front of your body is facing him. Your body is stationary and you're not going anywhere. Everything about your body language is saying "Stay where you are."

Instead, let your body say, "Here we go!" You will definitely need to use this move. Face the direction you want to go, use a happy voice and food, and move like you're late for a meeting. You're creating movement with your body language and your puppy will usually follow.

The following video shows moving a puppy away from something he finds interesting.

Video 6.23 How to effectively move a puppy away from something he's attracted to

Here is a real-life example. I (Marge) planned to meet a client at a quiet area of a local park. But that day the park service had a program for grade school children. They were right along the path. The puppy got excited by the children and the hawk the park ranger was holding.

Because I had chosen the location with care, there was plenty of space to increase distance from the kids and hawk and there were quieter areas in the park where we could work. First, we shortened the leash and moved the puppy away. Next, we let the puppy observe from a distance where he would not be disruptive. Then to get past the group to a quieter area, we increased distance some more, upped our rate of treats, and used our voices. Instead of practicing pulling toward the children and the action, the puppy was able to watch from a distance where he could be successful and then move on by.

Sometimes increasing distance isn't possible. Luckily, there are a few other things you can do.

3. Use better food. Harder work requires better pay. If I asked you to take the lid off my water bottle for $5, if you were able, you would probably do it. If I asked you to wash and clean my (very messy) car, inside and out for that amount, you'd take one look at my car and decline. The compensation doesn't match the effort. When you go out in public with your puppy, you'll need better food. It doesn't matter how much your puppy will work at home for kibble. New locations require better food. Most puppy owners underestimate the need for this. They bring boring food. Don't do that. Bring the roast chicken, beef, favorite cheese, or salmon dog food in a squeeze tube. Small bites, remember. But great food.

In the next video, when we move toward the traffic, we switch from regular treats to something special. You can see Cocoa notice the cars and then her owner offers her a camping food tube (see PuppySocialization.com/resources) filled with canned dog food.

Video 6.24 Cocoa succeeds in an intense environment with high value food

4. Increase your rate of reinforcement. Rate of reinforcement means how often you give treats. When walking your puppy at home, you might be able to go about ten steps between treats. When you're walking by a big distraction, or waiting while one moves by you, you might have to give your puppy a special treat more often. This could be every one or two steps if you're walking. The distractions increase the difficulty, so we reinforce more often.

5. Use your voice. It can help to use a happy voice when puppies (or even grown dogs) are startled or scared. "Yay! We love it when noisy trucks go by. That means I can feed you chicken!" Try not to continuously repeat "It's okay." Over time, that can come to mean something bad is about to happen. Also, try not to escalate by speaking louder and with more intensity. You end up sounding scary to your puppy, which makes the whole event more difficult for him.

6. Escape overly forward people. This is a special variation on increasing distance with movement. You'll almost certainly need to escape from well-meaning people at some point. This is super hard for most of us because we are breaking social expectations. Who wants to scoot away from some sweet children who want to pet the puppy? Their parents may have even prompted them to come over, so you'd be disappointing them, too. But you may have to do it, for your puppy's sake.

So it's good to prepare for this. Practice giving a firm "No" accompanied by a "stop" signal with your hand for those people who are coming too fast or won't respond to a polite request. You can also practice a polite phrase to say over your shoulder as you are leaving. Don't stop to explain; remember, the goal is to get your puppy out of the situation. If you allow a conversation to start, you're sunk. Believe us.

You can practice something like, "Sorry, maybe next time!" and be ready to use it as you slip away. But the most important part is stopping the person and getting out of there. We're serious about practicing. This video shows Marge helping a client work on stopping an oncoming, determined stranger.

Video 6.25 How to protect your dog from overly friendly people

7. If you can't turn the outing into a "win," pack it in. Sometimes we can't control a situation well enough to ensure positive outcomes. So leave. Don't risk something traumatic happening to your puppy because you don't want to hurt someone's feelings or you want to finish your latte. Leave. A single traumatic event— traumatic from your puppy's viewpoint—at the wrong time can impact him for the rest of his life. Don't risk it. Pack it in and leave.

Most puppies can weather a few mishaps just fine, especially if you react quickly to help. The following video shows Cocoa, the puppy afraid of cars and traffic, less than a year after we finished her puppy training. Her owners continued to work with Cocoa at a distance where she noticed cars but still felt safe. They got closer very slowly. They used food and play and continued their efforts through her first year of life. They did a fantastic job helping Cocoa blossom.

Video 6.26 Cocoa progressed well with practice and dedication from her owners

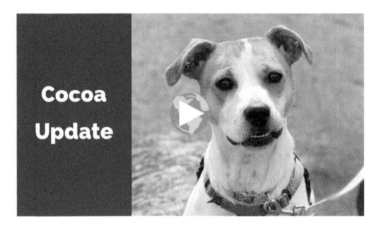

Puppy Classes

First, don't panic if you can't get to a puppy class, either because you live in a remote area, or the puppy classes available are not of the right type (see below), or because of the COVID-19 pandemic. If that's the case, follow the other guidelines in this book and do all the safe exposures to people, animals, and novelty you can. A good puppy class is wonderful, but your pup is not doomed to a life of fear without one.

But here's what a good puppy class can do. Well-run puppy classes give your puppy a chance to interact with and learn about members of his own species in a safe way. Puppy classes often have pups of various types: fluffy, smooth-coated, with prick ears, drop ears, raised tails, and more. And because these puppies are also in their SPS, they are more likely to interact with each other in a positive way.

The puppies in the adjacent photo are part of a puppy socialization and confidence group. Their owners are nearby with food and toys while both puppies explore novel objects and each other.

Look for a class specifically for puppies in their sensitive period for socialization. The cutoff for these classes is usually 16 weeks of age. The classes should follow the safety requirements listed in the AVSAB Position Statement on Puppy Socialization (American Veterinary Society of Animal Behavior, 2008a) and local health requirements related to the COVID-19 pandemic. (See the Position Statement, if you haven't checked it before, at PuppySocialization.com/resources.)

Classes for puppies in their SPS should focus on introducing them to a variety of objects, sounds, new people, and other puppies. This should be done with a liberal use of food and play. Good puppy classes can provide wonderful interactions. There should be no exercises for telling your puppy "No!" or correcting behavior. You want your puppy to learn to explore his world and you want him to be confident in doing so. Puppy class is a great opportunity to build positive associations and experiences with things like new people, new environments, other puppies, unstable surfaces, crutches and wheelchairs, and sometimes even some standard veterinary equipment like stethoscopes and otoscopes. A good instructor will also give you information about reading your puppy's body language and behavior. Remember, you shouldn't be socializing your puppy unless you can tell when he's relaxed and happy or a little bit (or a lot!) worried.

Puppies who are frightened should be given a safe place to hide during class. Usually, by watching the other puppies playing and having a good time, they will venture out on their own. Any effort to come out should be reinforced with food and/or play—whatever your puppy loves. Try not to lure your puppy out from his safe zone. Making him come out in order to eat won't make him feel better about leaving his hiding place. Let him come out on his own.

Puppy classes should also include some periods of free play in which puppies are allowed to play with each other. I (Marge) generally allow free play for 2–5-minute periods, if everything is going well. I interrupt play often, especially at the beginning. Owners take their puppies back to their mats to settle and rest before the next play interval. Puppies need to learn good play skills and one of those skills is learning to take little breaks in play. They are also learning to recover when excited and playing. As we mentioned previously, this is called an "off-switch" (McDevitt, 2007, p. 154). Trust me: it's a great skill for puppies and adult dogs to have.

If your puppy is comfortable, he should be exploring the class environment. Typical puppy behaviors include exploring, trying to get to other puppies or people, and playing with toys. These behaviors show interest in the environment. But a puppy who's hiding under your chair, taking a nap (a puppy who didn't have vaccinations that day), frantically trying to get you to pick him up, or bolting toward the door? It's likely the puppy is overwhelmed. The intensity of the experience is too great. That is sometimes referred to as flooding. Flooding is when an animal is trapped in an environment with scary things and can't escape. With flooding, the puppy is overwhelmed and typically either "shuts down" (becomes very passive and reluctant to move, acts sleepy, or moves in slow motion) or becomes frantic. Often, increasing distance from the group and putting up a physical barrier like a gate can help an overwhelmed puppy. Creating a place where he can watch from off to the side goes a long way toward helping the puppy feel safe so he can learn.

How to Find a Good Puppy Class

Here's what to look for in a puppy class.
- Unforced interaction with novel objects paired with food and/or play.
- Carefully supervised free play.
- Puppies allowed to hide when frightened.
- Practice with calming strategies (such as settling on a mat).
- Positive reinforcement using food and play.
- Information from the instructor about reading your puppy's body language.
- Reciprocal play, meaning the involved puppies are enjoying the interaction. One way to find out whether both puppies are enjoying the play is to separate them for 5–10 seconds. Release the quieter puppy first. If he re-engages with the more boisterous puppy, he's likely having a good time. If he walks away, he needs a break.

The following video shows lots of examples of things to look for in a class for puppies in their SPS. Marge coordinated the class. You can see the puppies experience wobbly surfaces, things to go under and through, a stethoscope, an otoscope, and more. Not shown in the video is that during the puppies' rest periods, Marge reviews information on body language, housetraining, puppy biting, socialization, and pairing food with new experiences.

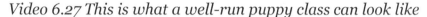

Video 6.27 This is what a well-run puppy class can look like

One of my (Eileen's) favorite things in this video is what happens after the 1:35 mark. The little chocolate Lab is checking out the skateboard and a big doodle comes and pushes her away and starts obnoxiously following her. (No, of course that's not my favorite part yet.) What happens next is that the owner of the bigger puppy quickly steps in and wrangles her pup away, following immediately with a treat. Then the Lab provides us with a wonderful example of puppy bounce-back. Her tail is down for about one second (at the 1:38–1:39 mark), as she runs away from the bigger pup, then it is up and wagging again as she returns. She checks out the big puppy's toy, passes through the same area, and continues exploring. No big deal!

So that's what a good puppy class looks like. Now we'll tell you what to avoid.

Signs You Should Avoid a Puppy Class

The things on the following list are more than "warning" signals. They are "danger" signals. If they are present, you should not take the class at all.

- Avoid classes that use special collars like choke chains, prong collars, or electronic or "stim" collars. These types of collars influence behavior by hurting or startling your puppy. These devices should never be used during a puppy's SPS (or, we would argue, any other time). Remember that we talked about dogs learning by association? If your puppy experiences pain or discomfort while he is learning about new things, he will likely form negative associations with those new things. I (Marge) have worked with many, many clients whose dogs' behavior worsened as a result of the use of those tools and methods.

- Avoid classes that use shake cans (cans filled with pebbles or pennies), air horns, compressed air, or verbal corrections like "Eh-eh" and "No!," used for punishment or interruption. While a simple "Eh-eh" or "No!" might seem harmless enough, your puppy doesn't speak in words. What makes those things seem effective initially is the implied threat behind them (your tone of voice and body language). Compressed air has recently gotten popular as a way to "interrupt" behavior in puppy class. But even on the website for the most popular brand of compressed air, it states, "Never use [this product] on young puppies. Early positive training is the best way to resolve puppy training issues." Puppy class should never include anything meant to startle or scare your puppy. You really don't want to startle or scare puppies during their SPS. Boy, could that backfire! I have, unfortunately, seen the results of these methods when people seek help later for their now-fearful puppy.

- Avoid classes that play "Pass the puppy" when the puppy is scared or has no choice. The old-fashioned "pass the puppy" game had owners pass their puppies to each other even if the puppies were struggling or afraid. Voluntary or lightly structured visits with new people are great as long as the puppy has a choice.
- Avoid classes where puppies are forced to interact with objects or people.
- Avoid classes where there is talk of dominance theory or pack leaders. This is outdated. Veterinarians who specialize in animal behavior issued a position statement against the use of dominance theory (American Veterinary Society of Animal Behavior, 2008b). This position statement is linked at PuppySocialization.com/resources.
- Avoid classes in which there is one-sided play between puppies or one puppy physically controlling other puppies repeatedly.
- Avoid classes where there is any use of force against a puppy, such as jerking on the leash or rolling him onto his back and pinning him (sometimes called an alpha roll). Both of these techniques are old-school dog training and harmful to puppies.
- Avoid classes that combine puppies of all ages and focus solely on "obedience." Unfortunately, some puppy classes we see offered include puppies in too large an age range: from very young puppies up to 6 months (26 weeks) old. Puppies at 8 and 26 weeks old are at different developmental stages. Also, the primary focus of some such classes is training behaviors, rather than taking advantage of the SPS and focusing on socialization.

If you are wondering what to do if your puppy is doing something you don't want him to (such as harassing another puppy), remember: you're the human and you have a leash and treats. If your puppy gets over-excited (or barky or afraid), move him farther away. Watch the video of Marge's class again and see how the doodle puppy's owner interrupted her puppy when it was being a little too much for the smaller Lab. Good puppy class instructors will help you do this if you don't think of it on your own. Increase distance to the point where your puppy can be successful.

What If You Enroll in a Puppy Class and They Do Things You Aren't Comfortable With?

This class is not for you. Leave.

We are so proud of you for noticing. You bought this book, you're reading and studying, and you've sensed a problem, even though you chose a class recommended by both your best friend and your veterinarian. You showed up at class and you observed the teacher pin a puppy to the ground. The teacher seemed friendly and authoritative. And she told the class with a smile on her face that the owner needed to "alpha roll" her puppy and show him who was the pack leader. But you know dominance theory is outdated and not practiced by trainers who stay up-to-date on current research. And you could clearly see that the puppy was scared.

We are giving you permission to leave. More than that, we are *advising* you to leave. Because even if you don't let the trainer touch your puppy (and some trainers don't ask permission; they may grab, knee, or slap dogs), many things can go wrong in such a class. If a trainer is using outdated methods like pinning a puppy to the ground, there is a good chance she won't be able to recognize the signs of fear and anxiety in puppies. That means if your puppy is fearful or a bit worried, the trainer will still likely proceed with whatever the plan is for that day. There's a chance she will blow an air horn or compressed air at another puppy and your puppy will hear it, too. She won't recognize that your puppy might need to start away from the group and work his way closer over time. She might not recognize that your puppy is fearful and won't be able to coach you through helping him.

Remember single-event learning? Negative experiences can be disastrous for your puppy. If the experience is intense enough, even a puppy in their SPS may not bounce back. The class will be hard on you, too, if you join and choose not to always follow the leader. I (Marge) have had clients tell me they were ridiculed and even shouted at. That's a lot of pressure. Just leave. No explanation required.

Puppies don't need discipline. They need advocates. They need teachers. They have so much to get used to in a human world.

Chapter 7. Fear and Special Challenges

There are always surprises. Eileen had no plan for a feral puppy to walk into her house and join her family. Some people have planned extensively for their puppy, only to figure out after they'd had him for a few days that he was deaf. Maybe you are raising a puppy for a special purpose and are wondering how to adapt this socialization stuff for him.

In many situations, the socialization adaptations will be minor. You may be able to figure them out on your own. But some situations will require professional help. We'll help you recognize signs that your puppy may need more than "normal" socialization.

Abnormal Fear During the SPS

Fear of novelty can be expected in an undersocialized adult dog. But such fear in a puppy in his SPS usually requires intervention.

To notice abnormal behavior, you need to know what's typical. And "typical" is hard to describe. The more I (Marge) train, the larger the variety of behaviors I encounter. I like to think of a bell curve for typical puppy behaviors. The majority of puppy behavior will fall into the big part of the bell, especially during the sensitive period for socialization. I expect most puppies to be inquisitive, social, playful, and to use their teeth. Of course, this will vary based on genetics and the puppy's environment before you brought him home. We expect variation.

But sometimes a puppy is an outlier. Most 10-week-old puppies don't bark and growl when they meet me. Most 10-week-old puppies don't cower away from me and refuse high-value food. Those behaviors are atypical. A 10-week-old puppy barking at strangers a block-and-a-half away won't "get used to people." He needs more help. Most puppies I encounter willingly take tidbits of tasty food and investigate new things placed in their environments. When a puppy does not do those things, the puppy might need more attention and training than a "typical" puppy. It doesn't mean your puppy won't turn out to be a fabulous dog or family pet. It can mean the behaviors you're seeing now will become more intense as your puppy grows up unless you intervene.

The 10-week-old puppy who barked and growled at me? I would expect the behavior to intensify when the puppy hits adolescence unless his owners work to change how he feels about the things that worry him. The healthy, hungry puppy who avoided me and would not eat food? I wouldn't be surprised if that puppy stayed fearful of new people when he got older. But those trajectories are not set in stone, especially after only one occurrence, if the owners get help.

The first thing I do when I encounter a puppy displaying overly fearful behavior in his SPS is refer the owners back to their veterinarian. Veterinarians are the team leaders for your puppy's physical and behavioral wellness. Your veterinarian may elect to bring a veterinary behavior specialist onto your puppy's team. Or she may refer you to a behavior consultant or work with you herself. Your veterinarian will also likely evaluate your puppy and any medical conditions that could be contributing to your puppy's behavior.

The following warning flags could indicate your puppy is fearful. Many of these match up with items on a list in the book *Control Unleashed: The Puppy Program* by

Leslie McDevitt (McDevitt, 2012, p. 43). McDevitt's book is linked at PuppySocialization.com/resources. We recommend that book for all new puppy owners. While this book is about socializing your puppy, McDevitt's book explains important skills to teach your puppy and how to teach them.

The following list contains some warning flags from a list in that book and a few of our own:

- You tend to describe him as "shy" or "nervous."
- His usual response to new things is to try to get away, whether it's by backing up, running, asking to be picked up, hiding in a corner, or any other way of escaping the new thing.
- He takes more than a few seconds to recover after being scared. Keep in mind the Lab puppy in the puppy class video. She recovered from a bigger dog scaring her in under two seconds.
- He has a hiding place where he commonly retreats.
- He growls or bites when you try to handle him.
- He's hard to handle at the vet, even after having happy visits.
- He stiffens, growls, or bites when someone approaches his food or toy.
- He runs away, trembles, drools, or pants when he hears a new or startling sound.

The good news is that there are some wonderful resources to help your puppy. Your veterinarian, a veterinary behaviorist and/or a skilled behavior consultant can often help turn things around. Again, the earlier the better.

"We thought he would grow out of it." I hear this from owners all the time. It's human nature to hope and believe things will fix themselves. But puppies don't grow out of being fearful of novel things. Things often get *worse* when the puppy hits adolescence. The average pet owner will need professional help for a puppy who is atypically fearful. Remember, puppies in the SPS are more likely to approach things, even things that scare them a little. So, a puppy in the SPS who you think of as "shy," "timid," or fearful is the opposite of typical and this is a behavioral emergency. Immediate intervention is needed for a couple of reasons. First, in the SPS, the puppy's temperament is more malleable than when he is older. Second, the emotions of fear generalize exceptionally well. Fear can keep animals safe from predators and is a strong survival behavior. But we don't want dogs to have that "Danger!" response anytime they encounter anything or anybody new!

The puppy in the adjacent photo is in his SPS. Does he look "relaxed and happy" or nervous and avoidant? If you guessed nervous, you're right. Notice how his mouth is clamped shut and his tail and paws are tucked under his body. It almost looks like he's trying to make himself look smaller, doesn't it? While he did show some improvement that day, he was not "relaxed and happy." His behavior was so

atypical for a puppy in his SPS, I (Marge) referred the owners to their veterinarian for a behavior evaluation. They made a behavior training plan for him that included working with me.

If you are raising your first puppy, or even your second or third, you may not have enough experience to judge whether your pup's responses are atypical. It's always good to contact a professional trainer, behavior consultant, or veterinary behaviorist if you have any concerns at all. The risks of not calling are much greater than the risk of calling if it turns out your puppy is fine. And yes, it does cost money to consult or hire a professional. But if the pup turns out to have problems, the longer you wait, the more intervention may be needed.

A Bad Day or a Fearful Puppy?

How can you tell if your puppy is fearful or just having a bad day?

Let's say you planned an outing for after your puppy wakes up from his morning nap. He did not have vaccinations that day or the day before. He did not have a busy morning. He's hungry, but not starving. You get to your location—a quiet spot in a park. But your puppy is not able to play, explore, or take food in this quiet location. Thinking back, you realize this is not the first time you noticed he was worried. There's a good chance he displayed some of those same body language cues when you first brought him home or at another location.

What can worried behavior look like? He might be frozen on the ground. He might be jumping on you and/or whining to be picked up. He might be trembling or scrambling to get back to your car. If your puppy does not begin behaving in a more typical puppy way (playing, exploring the quiet environment, interacting with familiar objects, eating, and drinking) after one or two visits to that location at quiet times, you need to act. You should seek guidance from a professional. If this happens once, the puppy might be having a sleepy day. But if it keeps happening, the sooner you get help from an expert, the better the outcomes for you and your puppy. Remember how short the SPS is. Sometimes the intervention of a professional can make all the difference, as it did for Star, one of the puppies featured in the "Special Cases" section below.

Fear Generalizes

One of the most important reasons for our puppies to have tons of positive experiences with people, animals, objects, and locations is to stay ahead of fear. Fear generalizes so easily. Achieving positive generalization can be a lot of work. But unfortunately, things can get categorized by your puppy as dangerous in a flash. Not only can they stay in that category persistently, but the fear can also spread. If a pup has a scary experience with, say, a man with a beard (sorry to keep picking on you guys!), the fear that event triggered can spread very fast. It may spread to other men with beards, to men who otherwise resemble the scary fellow, to men in general, to the location where the scare occurred, or even to other objects in the area. This is called *stimulus generalization* in behavior science textbooks (Mayer et al., 2019, p. 459).

How fearful or shy your puppy is will depend on lots of factors, including breed, the prenatal condition of the mother dog, the pup's place in the litter, health challenges, genetics, and more. Some of you will have puppies that are more fearful than others. But one thing we know: fear almost never goes away on its own. It's way easier to prevent fear from developing in the first place than it is to get rid of it.

Beating the Odds

Sometimes even dogs with lots of strikes against them can bounce back pretty well. Marge here. I had a client with a 10-week-old labradoodle puppy she had shipped from across the country. The puppy hid under the buffet in the dining room and barked at me for most of the first session, even when I was not interacting with him. The owner reported the dog didn't want to leave the house and definitely would not leave the yard. He also did not like loud noises, like trucks and lawn mowers.

The owner had two young children and a busy work and social life, sometimes with camera crews at her house. Based on my observations during our first session (even more than what I listed above), I recommended she return the puppy. He was not a good fit for their lifestyle. The owner told me she tried, but the breeder would not take him back. Plus, they loved him. The owner had the resources to pay for intensive training, so I came and worked with him multiple times a week for several weeks.

Happily, the puppy turned out great! I loved him. He had great focus, toy drive, could happily walk by loud machinery on the street, and welcome camera crews in the house. He would have made an awesome sports dog and I would have taken him in a minute. I say this not to toot my own horn, but to highlight the amount of work it took to help this puppy reach his full potential. Multiple sessions per week for several weeks with a skilled professional.

The moral of this story is that very little about behavior is written in stone. As I said before, dogs and their owners still surprise me. I know behavior is modifiable—within reason. The bigger questions are whether your goals for your puppy are modifiable and whether they are realistic for the puppy you have.

Fear vs. Phobia vs. Anxiety

We've mentioned that fear is a natural, necessary response. It is part of our behavioral palette (and that of dogs) because it can save our lives. The problems occur when it gets triggered repeatedly by a non-dangerous thing (as with phobias), or alternatively, when you just can't seem to turn it off (as with anxiety).

A PetMD article by veterinary behaviorist Dr. Wailani Sung (linked at PuppySocialization.com/resources) makes distinctions between fear, phobia, and anxiety. We quoted her definition of fear earlier in the book. Here are all three definitions together:

> • *Fear is the instinctual feeling of apprehension caused by a situation, person, or object that presents an external threat—whether it's real or perceived.*
>
> • *A phobia is the persistent and excessive fear of a specific stimulus.*
>
> • *Anxiety is the anticipation of unknown or imagined future dangers. This results in bodily reactions (known as physiologic reactions) that are normally associated with fear (Sung, 2019).*

Phobias and anxiety are actually medical diagnoses, and unless you are a veterinarian or a veterinary behaviorist, you can't make the diagnosis yourself. But you can learn the language and behaviors of fear and take note if your puppy stays afraid of things even after careful exposures. That's the time to consult these professionals, along with a dog behavior consultant or trainer who specializes in working with fearful dogs.

Veterinary behaviorist Dr. Karen Overall states the abnormally fearful or aggressive puppies need every intervention possible, and this can include medications (Overall, 2013, p. 129).

Special Cases

Every dog is unique. Smart trainers say every dog is a study of one. That means our focus needs to be on the puppy in front of us. Not what we expect the puppy to be, not what his littermates are like, or what his breed is expected to be like. Not even what he was like yesterday. We need to focus on the behavior of the puppy in the current moment.

Here are some common "special cases." These puppies actually turn out to need about the same things as the rest, sometimes with minor tweaks.

Playing Catch-up with Young or Adult Dogs Who Weren't Socialized During the SPS

We probably have convinced you by now that the sensitive period for socialization is a special time. Puppies are little sponges, and during this period they don't have a fear response to every novel thing they see. And if you do a lot of socialization—the right way, with exposures tied to good things—your puppy may generalize that new experiences are generally fun and cool. He'll at least know you've got his back and will help him handle new things.

We've also given you several examples of what happens when puppies aren't socialized during the SPS. Their default response to anything new—anything not part of their lives during the SPS—often becomes suspicion and fear. The new thing may be Grandma, whom you love with all your heart and assumed your dog would, too. She's family! But if your pup wasn't exposed to people other than the immediate family in your home during the SPS, your dog's response to Grandma might be fear. And that response could be the puppy's default to most people, other animals, noises—all the things we've mentioned.

This is not an absolute rule. As we've mentioned, some puppies, often from breeders who breed for solid temperament but sometimes by the luck of the draw, can bounce back from being undersocialized. But from the experiences of most trainers, these are the exceptions and not the rules.

So what do you do if you have a fearful dog who was insufficiently socialized as a pup and you are working on the same activities now? Trainers differ in the terminology—whether to call it socialization or not—since these dogs are past their SPS. The two of us don't call it socialization. But the action is the same. You do what we've told you to do in this book, but it will go way, way slower. Because now the dog's default response is not happy curiosity or even neutral. It's likely fear. And the help of a skilled professional is probably necessary.

Star's Story: It Wasn't Too Late after All

Star is an adorable Sheltie puppy whom I (Marge) met when she was 16 weeks old. Star was the first "pandemic puppy" I worked with, meaning she was brought home during the COVID-19 pandemic while her owners were sheltering in place. As a result, she did not meet many people outside her immediate family. Nor did she go to any new locations or encounter much novelty. She was loved and played with and went for hikes in the woods daily, but that was the extent of her activity during the lockdown.

When I first met Star, we met outside in her front yard. She was wary of me and reluctant to approach. I used quiet body language and tossed treats behind Star. After about 35 minutes, she felt safe enough to take treats from my hand. Over several sessions, we began introducing novelty (items she was familiar with but in a new location) to Star. While she was cautious at first, Star grew to enjoy our games with new things. Star's owner was amazing and allowed Star to progress at her own pace. She also learned to have very good timing when pairing new experiences with food, and she was generous with food and play. She was exactly what Star needed. And Star blossomed.

You can see in the following video how Star went from a puppy who was afraid of the movement of a water bottle to a confident little girl who can have fun and play in new environments.

Video 7.1 Star's wonderful progress

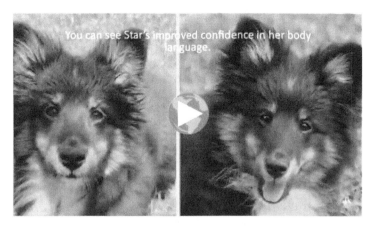

Dogs and their owners surprise me all the time. Star and her owner surprised me in the best way: by exceeding my expectations. Even though Star was likely out of her sensitive period for socialization when we met, she made exceptional progress with concerted and focused help from her owner, with guidance from a professional behavior consultant. While we may never know, it's a good bet Star's progress would likely have been much slower, and required much more work, if her owner had sought help a month or two later. So if your puppy is like Star, whether he's in or out of his SPS, don't wait. Get help today.

Feral Puppies

I (Eileen) took in a feral puppy whose SPS was just ending. I got into her "safe" category; no other humans did. She was born and grew up in the woods and was probably affected by the malnutrition of her mother and prenatal stress. At 10 weeks, she was that outlier puppy whom Marge described above: she growled at and even advanced on people she didn't know. What did Clara's training consist of? Exactly what we describe here, but at 1/100th of the normal pace. That's not an exaggeration, by the way. We did the math. Also, I didn't do it on my own. It took steady, long-term work with a skilled professional to help Clara approach normalcy around people.

Clara at nine years old now has a few more human friends. She does better than many "normal" dogs at the vet. She can walk happily on leash in public parks and shopping centers, even among crowds. She is relaxed and generally happy unless someone approaches her straight-on with eye contact, wanting to interact. Then she will bark and dodge away. It's my job to make sure people don't approach her. She can appear so normal, since she is curious and enjoys being out and about. But she could never be a competition dog or a therapy dog or even go to a family gathering. If anyone visits my home, Clara stays in a back room with a series of food toys and frequent visits from me.

The important thing to remember is that we just can't predict how well a puppy with an atypical start in life can do. Some feral puppies transition all the way into the human world and you would never know that they started out wild. Some are like Clara, who made spectacular progress in all other ways but could never accept that humans in general are safe. Some do fine in their home environment but nowhere else. What we do know is that you will have a lot of extra work. This book will help you, but Marge and I both hope you will enlist the help of a professional.

Puppies Who Are Blind or Deaf

Puppies born blind or deaf don't know they are "missing" a sense. They usually get along just fine. Like other puppies, they still need exposure to the things they are going to encounter during the course of their lifetime. Most people will need some help from a trainer to learn how best to communicate with and train a puppy who is blind, deaf, or both.

Marge here. The following video is from the first time Luna and her littermate, Marlo, came to me for training. Luna is completely deaf and almost completely blind, but I did almost nothing different with her socialization than I would do with a sighted and hearing puppy.

This video shows what "starting small" can look like for a dog with sensory impairments. Luna is in a new location and has never met me and you can bet I am pairing my own presence with some good treats.

Her owners had already taught her to go up stairs. You can see her lower her chin to feel the next step. Luna is learning a rule structure for playing tug. She sits, then a tap on the chest tells her she can start the game. Then she gets a treat for releasing the tug. Associating this fun game (and look how fast she is learning!) with going somewhere new is helping Luna create good associations to new experiences.

Video 7.2 Luna has fun practicing tug with clear rules in a new environment

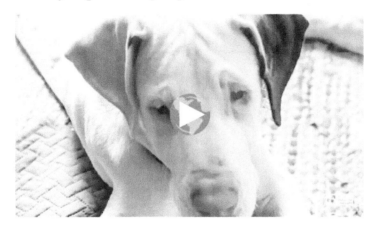

Service Dogs

Many people wonder whether they should socialize their puppy who is a candidate for service dog work. They don't want him to interact with people during future public work.

Yes, future service dogs need to be socialized! We hope we've convinced you by now that all puppies need to form positive associations with the things they will encounter during their lifetime. This couldn't be more true than for service dogs. Remember when we said a dog's first reaction to something they weren't exposed to during their SPS will most likely be fear? Fearful dogs can't do their jobs. They are worried for their safety.

Socialization is so important for this category of dogs that service dog programs will often send their puppies out to foster families for their first year for socialization. Every puppy needs socialization. A dog who grows up to be "overly" friendly can generally be trained appropriate manners and service dog behaviors. A dog who is fearful can't do service work in public at all.

Chapter 8. What's Next?

Congratulations! You've learned what puppy socialization is and how to do it. And you've learned to recognize when dogs are relaxed and happy or worried and avoidant. Finally, through practicing the pairing exercises, you've developed the skills to help build confidence in your puppy. Socialization may seem like a lot of work, but trust us when we say it will pay off for years to come. You won't leave those skills behind when your puppy leaves the SPS and becomes a juvenile. You'll carry those new skills with you throughout your dog's life.

Are you thinking all the heavy lifting is done and now you can coast through the rest of your dog's puppyhood? Not so fast. The SPS is only the beginning of what your puppy needs from you. Controlled exposures and socialization-type activities should continue through your puppy's first year of life. If there is any particular thing or category your puppy still seems a little worried about, you'll need to continue controlled and positive exposures. Your puppy will not grow out of his puppy worries or fears. They will most likely intensify when he reaches adolescence.

If you've followed the guidelines in this book, and you still have a puppy who you feel is shy, timid, or lacking confidence, please seek help. We have included contact lists for accredited trainers, behavior consultants, and veterinary behaviorists in our resource list at PuppySocialization.com/resources.

If you've got a friendly, confident puppy, congratulations again. You may have gotten a head start with genetics, but you put in the work. And take a look at the change in your own behavior. You can read your dog. You likely make thoughtful preparations when you take him anywhere. You can tell when he needs help. And you know how to give it to him.

Take a moment and congratulate yourself for taking the time to learn about your puppy's behavioral development. It is awe-inspiring if you think about it. We welcome an animal of a different species into our home and life. They become our friends, our confidantes, and if we're lucky and open to it, our teachers. As we, as a species, learn more about teaching and behavior, it seems like all of a sudden dogs are capable of so much more. But really, they had it all along.

You are well on your way to being a compassionate teacher for your puppy.

*****

Thank you for reading our book. We hope you will review it wherever you made your purchase. Reviews help more readers like you find us.

Acknowledgements

So many people gave help and encouragement for this book! So thank you first to all the people, too many to name, who said, "Please create this book! We need it!" We hope we have lived up to your faith in us.

Joyce Loebig did an outstanding job copyediting (plus some crucial structural editing) and we can't thank her enough for her expertise, fine eye, and patience. Much of what is good about the book is because of Joyce, who managed to copyedit and make big-picture suggestions on one fantastic read-through. We want to make it clear that we worked on the book for three months after Joyce had it, so anything that is not according to good style practices and any actual errors were undoubtedly created by us after Joyce edited.

Dr. Karen London's help, support, enthusiasm have been invaluable. She was kind enough to read for specific issues and in addition helped guide us away from some major faux pas. As with Joyce, we tinkered with the book after Karen went through it, so anything awkward is on us.

Thank you so much to our pre-readers: Leslie McDevitt, MLA, CDBC, CPDT-KA; Alanna Lowry, DVM; Karen London, PhD, CAAB, CPDT-KA; Cathy Kreis, DVM; and Sara Bennett, DVM, MS, DACVB. It is a big task to read a book for the purpose of writing about it, and we appreciate their time and kind words so much!

We both appreciate Marge's clients so much for allowing us to use footage of them and their wonderful puppies. Marge often says she learns from each interaction with her clients and their dogs. She is grateful and humbled by her clients' trust in her and their contributions to her learning. They are the reason there is a book.

Last and most important, since the book would never have happened without their unflagging support, thank you to Marge's husband Bob Rogers and Eileen's partner Ruth Byrn. Bob also proofread, and here are our thanks for that often-thankless task! Ruth, an accomplished writer herself, helped Eileen throughout the process to convey ideas so they can be understood by a wide readership and provided a sounding board for her endless thoughts and questions about the book.

About the Authors

Marge Rogers, CBCC-KA, CPDT-KA, CCUI,
Certified Fear Free Professional

Marge Rogers is a certified professional dog trainer and canine behavior consultant. She has thousands of hours teaching owners, dogs, and other canine professionals.

One of the things you will notice right away about Marge is her enthusiasm for sharing information about training and behavior. Marge changed her life and career when she learned about the science of behavior. She is warm and engaging when sharing what she learned with her clients and other canine professionals.

Dog training and behavior aren't rocket science. They are behavioral science. Marge boils down the science into relatable, easy-to-understand information. Even her business name is educational: Rewarded Behavior Continues (rewardedbehaviorcontinues.com).

Marge earned certifications in training and behavior consulting (canine) through the Certification Council for Professional Dog Trainers. She is also a Certified Control Unleashed® Instructor and a Certified Fear Free® Professional. She received additional training on child-dog safety and family dynamics through Family Paws Parent Education. Marge is an engaging, experienced seminar presenter and has offered continuing education workshops for veterinarians and veterinary technicians in NC. She has worked with corporations (CSX, Babies 'R Us), communities, and schools to bust myths about dogs and dog training.

About the Authors

Eileen Anderson, BM, MM, MS

.

Eileen Anderson writes about her life with multiple dogs with a focus on describing positive reinforcement-based training to pet owners and beginner trainers. Her well-known blog, eileenanddogs.com, has been featured on Freshly Pressed by Wordpress.com and won the award, "The Academy Applauds" in 2014 from The Academy for Dog Trainers. She published a book on canine cognitive dysfunction in 2016 that won a Maxwell Award from the Dog Writers Association of America.

She has written for *Clean Run, Whole Dog Journal*, the *IAABC Journal*, and *BARKS from the Guild*. Her articles, training videos, and photos of dog body language have been incorporated into curricula worldwide. She holds an eight-week, biannual writing mentorship through the IAABC. She also works as a freelance editor and mentor for writers.

Eileen has worked professionally as a writer and academic editor, an orchestral musician, a network administrator, a remedial college math instructor, a bookkeeper, a social work caseworker, and a trainer of computer skills in academic and workplace settings. She has co-written successful grant applications totaling $5 million. She has a long-standing interest in making technology accessible to women, people with limited literacy skills, and other underserved populations. She holds bachelor's and master's degrees in music performance and a master's degree in engineering science.

Media Credits and Permissions

The photo of the border collie puppy (Flynn) in "Note from the Authors" and the two photos of the border collie puppy (Pip) in "How to Greet a Dog" are copyright Alanna Lowry.

All other images are copyright either Marge Rogers or Eileen Anderson.

All media portraying Marge's clients are published with signed releases.

References

American Psychology Association. (2021) *APA Dictionary of Psychology Online.* Retrieved May 19, 2021. https://dictionary.apa.org/approach-avoidance-conflict

American Veterinary Society of Animal Behavior. (2007). AVSAB Position Statement: The Use of Punishment for Behavior Modification in Animals. (currently offline)

American Veterinary Society of Animal Behavior. (2008a). *AVSAB Position Statement on Puppy Socialization.*
https://avsab.org/wp-content/uploads/2018/03/Puppy_Socialization_Position_Statement_Download_-_10-3-14.pdf

American Veterinary Society of Animal Behavior. (2008b). *AVSAB Position Statement on the Use of Dominance Theory in Behavior Modification of Animals.* https://avsab.org/wp-content/uploads/2018/03/Dominance_Position_Statement_download-10-3-14.pdf

Arai, S., Ohtani, N., & Ohta, M. (2011). Importance of bringing dogs in contact with children during their socialization period for better behavior. *Journal of Veterinary Medical Science*, 1101210440–1101210440.

Bateson, P. (1979). How do sensitive periods arise and what are they for?. *Animal Behaviour, 27*, 470-486.

Borchelt, P. L. (1983). Aggressive behavior of dogs kept as companion animals: Classification and influence of sex, reproductive status and breed. *Applied Animal Ethology, 10*(1–2), 45–61.

Bradshaw, J. W., Pullen, A. J., & Rooney, N. J. (2015). Why do adult dogs 'play'? *Behavioural Processes, 110*, 82–87.

Calder, C. (2020). "Fear and Fear-related Aggression in Dogs." Dr. Sophia Yin. https://perma.cc/S9HK-UJJR

Centers for Disease Control. (2001). *Nonfatal Dog Bite—Related Injuries Treated in Hospital Emergency Departments—United States, 2001.*
https://www.cdc.gov/mmwr/preview/mmwrhtml/mm5226a1.htm

Colombo, J. (1982). The critical period concept: Research, methodology, and theoretical issues. *Psychological Bulletin, 91*(2), 260.

Coppinger, R., & Coppinger, L. (2002). *Dogs: A new understanding of canine origin, behavior and evolution.* University of Chicago Press.

Domjan, M. (2018). *The essentials of conditioning and learning*. American Psychological Association.

Donaldson, J. (2013). *Culture Clash*. Dogwise Publishing.

Dunbar, I. (2001a). *AFTER You Get Your Puppy*. James & Kenneth Publishers.

Dunbar, I. (2001b). *BEFORE You Get Your Puppy*. James & Kenneth Publishers.

Fox, M. W., Inman, O., & Glisson, S. (1968). Age differences in central nervous effects of visual deprivation in the dog. *Developmental Psychobiology: The Journal of the International Society for Developmental Psychobiology*, *1*(1), 48–54.

Fox, M. W., & Stelzner, D. (1966). Behavioural effects of differential early experience in the dog. *Animal behaviour*, *14*(2-3), 273-281.

Grier, K. C. (2010). *Pets in America: A history*. UNC Press Books.

Hammerle, M., Horst, C., Levine, E., Overall, K., Radosta, L., Rafter-Ritchie, M., & Yin, S. (2015). *2015 AAHA canine and feline behavior management guidelines: Age and Behavior*. American Animal Hospital Association. https://www.aaha.org/aaha-guidelines/behavior-management/age-and-behavior/

Haug, L. I. (2008). Canine aggression toward unfamiliar people and dogs. *Veterinary Clinics of North America: Small Animal Practice*, *38*(5), 1023–1041.

Held, S. D., & Špinka, M. (2011). Animal play and animal welfare. *Animal Behaviour*, *81*(5), 891–899.

Houpt, K. A. (2007). Genetics of canine behavior. *Acta Veterinaria Brno*, *76*(3), 431–444. https://doi.org/10.2754/avb200776030431

Kalnajs, S. (2006). *The language of dogs*. Blue Dog Training and Behavior.

Killion, J. (2014, December 5). *PUPPY CULTURE FILM* [DVD]. Madcap Productions.

Knudsen, E. I. (2004). Sensitive periods in the development of the brain and behavior. *Journal of cognitive neuroscience*, *16*(8), 1412-1425.

Korbelik, J., Rand, J. S., & Morton, J. M. (2011). Comparison of early socialization practices used for litters of small-scale registered dog breeders and nonregistered dog breeders. *Journal of the American Veterinary Medical Association*, *239*(8), 1090–1097. https://doi.org/10.2460/javma.239.8.1090

Lindsay, S. R. (2001). Handbook of applied dog behavior and training. Vol. 2, Etiology and assessment of behavior problems. Iowa State University Press.

Martin, K. M., & Martin, D. (2011). *Puppy Start Right: Foundation Training for the Companion Dog*. Karen Pryor Clickertraining.

Mayer, G. R., Sulzer-Azaroff, B., & Wallace, M. (2019). *Behavior Analysis for Lasting Change* (4th ed.). Sloan Publishing.

McDevitt, L. (2007). *Control unleashed: Creating a focused and confident dog.* Clean Run Productions LLC.

McDevitt, L. (2012). *Control unleashed: The puppy program.* Clean Run Productions, LLC.

Merriam-Webster. (n.d.). Socialization. In *Merriam-Webster.com Dictionary.* Retrieved June 1, 2019, from

https://www.merriam-webster.com/dictionary/socialization

Montessori, M., & Carter, B. (1936). *The secret of childhood.* Calcutta: Orient Longmans.

Overall, K. L. (2004, August 31). Storm phobias. *DVM 360.*

https://www.dvm360.com/view/storm-phobias

Overall, K. L. (2013). Manual of clinical behavioral medicine for dogs and cats.

Reid, P. J. (1996). Excel-erated learning: Explaining in plain English how dogs learn and how best to teach them. James & Kenneth Publishers.

Scott, J. P., & Fuller, J. L. (2012). *Genetics and the Social Behaviour of the Dog.* University of Chicago Press.

Serpell, J. (Ed.). (2017). The domestic dog: Its evolution, behavior and interactions with people (2nd ed.). Cambridge University Press.

Siracusa, C., Provoost, L., & Reisner, I. R. (2017). Dog- and owner-related risk factors for consideration of euthanasia or rehoming before a referral behavioral consultation and for euthanizing or rehoming the dog after the consultation. *Journal of Veterinary Behavior, 22,* 46–56.

https://doi.org/10.1016/j.jveb.2017.09.007

Sung, W. (2019). Extreme Fear and Anxiety in Dogs. *PetMD.* https://www.petmd.com/dog/conditions/behavioral/c_dg_fears_phobia_anxiety

Tiira, K., Sulkama, S., & Lohi, H. (2016). Prevalence, comorbidity, and behavioral variation in canine anxiety. *Journal of Veterinary Behavior, 16,* 36–44.

Tóth, L., Gácsi, M., Topál, J., & Miklósi, A. (2008). Playing styles and possible causative factors in dogs' behaviour when playing with humans. *Applied Animal Behaviour Science, 114*(3-4), 473-484.

Vegh, E., & Bertilsson, E. (2012). *Agility right from the start: The ultimate training guide to America's fastest-growing dog sport.* Karen Pryor Clickertraining.

Von Pfeil, D. J., & DeCamp, C. E. (2009). The epiphyseal plate: physiology, anatomy, and trauma. *Compendium (Yardley, PA), 31*(8), E1-11.

Yin, S. (2011). *Perfect Puppy in 7 Days: How to Start Your Puppy Off Right.* Cattledog Publishing, 2011.

Made in the USA
Las Vegas, NV
05 December 2023

82136249R00136